Hands-on Exercise Manual for

LabVIEW® Programming, Data Acquisition and Analysis

ISBN 0-13-030368-2

NATIONAL INSTRUMENTS | VIRTUAL INSTRUMENTATION SERIES

Mahesh L. Chugani • Abhay R. Samant • Michael Cerna
■ **LabVIEW Signal Processing**

Rahman Jamal • Herbert Pichlik
■ **LabVIEW Applications**

Shahid Khalid
■ **LabWindows/CVI Programming For Beginners**

Barry Paton
■ **Sensors, Transducers & LabVIEW**

Jeffrey Travis
■ **Internet Applications in LabVIEW**

Lisa K. Wells • Jeffrey Travis
■ **LabVIEW For Everyone**

Hands-on Exercise Manual for

LabVIEW® Programming, Data Acquisition and Analysis

▲ Jeffrey Y. Beyon

Prentice Hall PTR
Upper Saddle River, New Jersey 07458
www.phptr.com

Library of Congress Cataloging-in-Publication Data Available

Acquisitions Editor: *Bernard Goodwin*
Production Editor: *Rose Kernan*
Cover Design: *Nina Scuderi*
Cover Design Director: *Jerry Votta*
Manufacturing Manager: *Alexis R. Heydi*

 © 2001 Prentice Hall PTR
Prentice-Hall, Inc.
Upper Saddle River, NJ 07458

Prentice Hall books are widely used by corporations and government agencies for training, marketing, and resale.

The publisher offers discounts on this book when ordered in bulk quantities.
For more information, contact Corporate Sales Department, phone: 800-382-3419;
fax: 201-236-7141; e-mail: corpsales@prenhall.com
or write: Prentice Hall PTR
  Corporate Sales Department
  One Lake Street
  Upper Saddle River, NJ 07458

All rights reserved. No part of this book may be reproduced, in any form or by any means, without permission in writing from the publisher.

All product names mentioned herein are the trademarks or registered trademarks of their respective owners.

Printed in the United States of America
10 9 8 7 6 5 4 3 2 1

ISBN 0-13-030368-2

Prentice-Hall International (UK) Limited, **London**
Prentice-Hall of Australia Pty. Limited, **Sydney**
Prentice-Hall Canada Inc., **Toronto**
Prentice-Hall Hispanoamericana, S.A., **Mexico**
Prentice-Hall of India Private Limited, **New Delhi**
Prentice-Hall of Japan, Inc., **Tokyo**
Pearson Education Asia Pte. Ltd.
Editora Prentice-Hall do Brasil, Ltda., **Rio de Janeiro**

Contents

Preface   ix

**Guidelines for
the Reader**   xi

**Drill Problems
Day 1**   1

    Drill Problem 1.1   p021_WirePrac.vi 2
    Drill Problem 1.2   p022_WirePrac.vi 3
    Drill Problem 1.3   p023_ShrtCutPrac.vi 4
    Drill Problem 1.4   p031_dBcalc.vi 4
    Drill Problem 1.5   p032_QuitPrompt.vi 5
    Drill Problem 1.6   p033_RNDisplay.v 7

Drill Problem 1.7	p041_ForLp&Indxng.vi	9
Drill Problem 1.8	p042_WhileLp&ShftReg.vi	11
Drill Problem 1.9	p043_Loops&Conditions.vi	13

Drill Problems Day 2   17

Drill Problem 2.1	s051_MechActionOfBooleans.vi	18
Drill Problem 2.2	p052_MultiChChartGraph.vi	18
Drill Problem 2.3	p053_MultiChGphXo.vi	19
Drill Problem 2.4	p054_MultiChXYGph.vi	21
Drill Problem 2.5	p061_ArryPrac.vi	23
Drill Problem 2.6	p062_ClustrFcnPrac.vi	24
Drill Problem 2.7	p063_IntstyGphPxlByPxl.vi	26
Drill Problem 2.8	p064_IntstyGphAttNode.vi	28
Drill Problem 2.9	p065_Arry&ClustrUpdPrac.vi	29

Drill Problems Day 3   31

Drill Problem 3.1	s081_AcqNScans.vi	32
Drill Problem 3.2	s082_CtsAcq&Chart(buff).vi	32
Drill Problem 3.3	s091_Gen1PtOn1Chwith_s082.vi	33
Drill Problem 3.4	s092_FGwith_s082.vi	34
Drill Problem 3.5	s101_CtsPTrn(8253)with_s082.vi	34
Drill Problem 3.6	s102_CntEvnts(8253).vi	35
Drill Problem 3.7	p111_WrtRdBin1D.vi	35
Drill Problem 3.8	p112_WrtRdASCII.vi	37
Drill Problem 3.9	p121_TypeCast.vi	38
Drill Problem 3.10	p141_AliasTest.vi	
	p141_FFTproc.vi	42

Drill Problems
Day 4   47

Drill Problem 4.1	s081_AcqNScansDTrig.vi	48
Drill Problem 4.2	s082_AcqNScansExtChClk.vi	48
Drill Problem 4.3	s083_ContAcq&GphExtScanClk.vi	49
Drill Problem 4.4	p101_DigWrt&AI.vi	
	p101_AISmplChs.vi	50
Drill Problem 4.5	p111_WrtRdBin1DCts.vi	53
Drill Problem 4.6	p112_WrtRdBin2D.vi	54
Drill Problem 4.7	p113_WrtRdASCIICts.vi	56
Drill Problem 4.8	p114_WrtRdMixed.vi	58
Drill Problem 4.9	p115_WrtRdDatalog.vi	60
Drill Problem 4.10	p116_RtrvDtalogDtaHalo.vi	61
Drill Problem 4.11	p117_SvDataWithPref.vi	62
Drill Problem 4.12	p121_ScrllbarCtrl.vi	65
Drill Problem 4.13	s131_LV&Serl.vi	66
Drill Problem 4.14	s132_LV&GPIB.vi	67
Drill Problem 4.15	p141_AcqNScns1Ch_QS1D&FFT.vi	67

Drill Problems
Day 5   69

Drill Problem 5.1	s141_FreqResp.vi	70
Drill Problem 5.2	p142_FFTPairTest.vi	70
Drill Problem 5.3	s143_DecIntTest.vi	72
Drill Problem 5.4	s144_AliasTest&LPF.vi	72
Drill Problem 5.5	p145_FittingPrac.vi	73
Drill Problem 5.6	p151_While&Occrnce.vi	
	p151_WhileWoOccrnce.vi	74
Drill Problem 5.7	p152_ErrHndler_Main.vi	

p152_ErrHndler_Proc1.vi
p152_ErrHndler_Proc2.vi
p152_ErrHndler_Proc3.vi
s152_ErrHndler_ErrDiply.vi 76

Drill Problems
Day 6   81

Drill Problem 6.1	p151_ListVIs.vi	82
Drill Problem 6.2	p152_DynamicLoad.vi	83
Drill Problem 6.3	p153_init.vi	
	p153_main.vi	
	p153_menu1.vi	
	p153_menu2.vi	85
Drill Problem 6.4	p154_StatCheck.vi	90
Drill Problem 6.5	p155_4ToggleSW.vi	91
Drill Problem 6.6	p156_LocalVar&SR.vi	
	p156_AcquireData.vi	
	p156_ProcData.vi	93
Drill Problem 6.7	s156_LocalVar&SR.vi	95

Index   99

Preface

The first edition of *Hands-on Exercise Manual for LabVIEW Programming, Data Acquisition and Analysis* has been written to serve as a supplementary exercise manual for the main text *LabVIEW Programming, Data Acquisition and Analysis*. Originally, this manual started as a main text for the biannual G-programming workshop that I have been offering to the engineers and researchers in the local community. This first edition is the result of adding more examples and making a few corrections to the original manual.

Without practicing, the learning cycle of any subject can never be completed, and the ultimate goal of this exercise manual is to complete the cycle. All of the examples are carefully chosen to offer maximum efficiency in learning the G-programming language with LabVIEW in a minimal amount of time. For readers who are anxious to implement any example in their application immediately, solutions to all of the problems are provided on the accompanying CD-ROM. The CD-ROM also contains an evaluation copy of LabVIEW so that readers can create some simple VIs and execute them without having the full version of LabVIEW.

Each example in this manual is carefully designed so that readers can use it as a starting point in their application. In order to achieve such a task, the following rules have been abided by at all times during the design process of each example:

1. **Keep each example simple**. A formidable-looking example would be nothing but a waste of time for both readers and myself. It can deliver some lessons, but going through other people's codes (in any programming language) to learn the programming language is always the worst way of learning it. Readers should first understand the basics, practice on some simple exercises, then *write their own programs*. Also, readers generally will not even look at such a complicated example. However, if each example is kept simple but meaningful, readers can easily understand the basic concept so that they can *extend* it to their applications easily. The fact is that eventually *the readers* are the one who will be writing their own applications. So all of the examples in this manual are kept simple, but concise and meaningful.
2. **Keep each example practical**. This manual has not been written to show every feature that LabVIEW has. Yet it is written for both beginners and advanced LabVIEW programmers. Beginners will find many examples ready to be implemented in their applications. Advanced programmers will find many examples to be a good chance to update their programming technique. Most of the examples in this manual originated from the actual applications that I have written for private companies and engineers, so it should not be too difficult for readers to agree with me on the practical aspects of the examples in this manual.

I would like to thank the many people who helped me complete this manual. My thanks to Dr. Siva M. Mangalam of Tao Systems, for the inspiration of many practical examples, especially those in the last section, Day 6. I am also grateful to my colleagues at Christopher Newport University, who helped make the biannual G-programming workshop a series of successes. Above all, special thanks to my family, Mina and Jeffrey, Jr., for their support and understanding during the never-ending days and nights spent completing this project.

Guidelines for the Readers

Use of This Manual

There are six sessions in this manual: Day 1, Day 2, Day 3, Day 4, Day 5, and Day 6. Each session has many drill problems with step-by-step instructions. Each drill problem has six subtitles. The following is an example:

1. VIs to be used: **p021_WirePrac.vi** (Template provided.)
2. Objective: To learn the correct and accurate wiring technique.
3. Estimated time: 10–15 minutes
4. Related chapter: Chapter 2
5. Key objects, VIs, and functions in this drill problem:
   Functions >> Data Acquisition >> Analog Input >> AI Config.vi
6. Instructions

The first subtitle, "VIs to be used," lists all of the VIs that are used in the corresponding drill problem. If the VI is a template, and you will have to complete it, it will state "(Template provided.)." If the VI is already complete and ready to execute, it will state "(The VI is already complete.)." If neither the VI is complete nor a template VI is provided, it will state "(You need to create a new VI.)" If the label starts with the letter **s**, the VI is already complete; otherwise, it is a template. For example, **p021_WirePrac.vi** is a template, whereas **s021_WirePrac.vi** is the solution. All of the solution VIs are located in the folders **Day1Soln** through **Day6Soln** on the accompanying CD-ROM.

The second subtitle, "Objective," addresses the goal of the drill problem briefly. The third subtitle, "Estimated time," is an approximation of the time that you may need to complete the problem. The fourth subtitle, "Related chapter," indicates the chapter where you can find the related topics in the main text. The fifth subtitle, "Key objects, VIs, and ... ," lists the labels of controls, indicators, VIs, and functions as well as their path to help you find them easily. The sixth subtitle, "Instructions," presents detailed descriptions about the drill problem as well as step-by-step instructions to complete the VI(s).

The best way to use this manual is by reviewing both the main text and this manual together. For example, after finishing each chapter of the main text, try the drill problems that correspond to that chapter. However, working on the drill problems without the main text is also possible since each exercise provides detailed steps to create the VI and explains its functionality. If you are an experienced LabVIEW programmer, you can definitely start with this manual. If you are new to LabVIEW, reviewing both the main text and this manual in parallel is recommended.

Equipment Recommended, but Not Required, for the Exercises in This Manual

- LabPC-1200 or any data acquisition board by National Instruments and proper cables
- CB-50 or any proper terminal block by National Instruments
- Any GPIB board by National Instruments and proper cables
- Screwdriver
- Jumper wires
- Function generator and appropriate cable connections
- 1.5 V Battery of any size and a battery socket

Accompanying CD-ROM

The accompanying CD-ROM provides readers with the template VIs of the exercise problems in this manual and the complete solutions to them. Also included is an evaluation copy of LabVIEW with limited functionality. Most of the problems may be completed using the evaluation copy, but some will require the full version of LabVIEW.

Installation of Template and Solution VIs

The following steps will allow you to install the complete set of the template and the solution VIs in LabVIEW for easy access to them. Those steps, however, are applicable to the full version of LabVIEW only. You should manually find and open those VIs if the evaluation copy of LabVIEW is used.

1. Start the full version of LabVIEW if it is not running already.
2. Go to the pull-down menu **Edit** and choose **Edit Control & Function Palettes …** . This will bring up the **Controls** and the **Functions** palettes.
3. Go to **Functions**, and pin down the subpalette **User Libraries**.
4. Right click (PC platforms), or **Command**-click (Macintosh platforms) in any empty space in **User Libraries**. Choose **Insert >> Submenu …** . (The symbol >> indicates the path. See the next section **Conventions** for the complete list of conventions used in this manual.) This will bring up a window as shown:

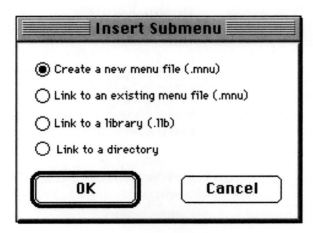

5. Select the second option **Link to an existing menu file (.mnu)**. This will bring up a directory navigation window.
6. Find the file **dir.mnu** in the directory where all of the six example folders are located. (Day1VIs, Day2VIs, Day3VIs, Day4VIs, Day5VIs, and Day6VIs) It is recommended that you copy the six example folders onto your hard drive, and use them for the problems.
7. Once you select **dir.mnu** in Step 6, you will see a new subpalette **DrillProblems**. Choose **Save Changes** to save the changes. Now, you can easily access all of the drill problem VIs in **Functions** >> **User Libraries** >> **DrillProblems**.
8. Go to the **Edit** pull-down menu, and select **Preferences**.
9. Select the menu **Block Diagram**.
10. Check **Show dots at wire junctions**.
11. Select the menu **Front Panel**.
12. Uncheck **End text entry with Return key (same as Enter key)**.
13. Now, you are ready to begin the drill problems.

Conventions

The following summarizes the conventions used in this manual:

Convention	Definition
bold	directories; folders; file names; object labels; file extensions; functions and item names of LabVIEW; menu selections; option selections; vectors; matrices
	e.g., Save **test.vi** in the folder **My Project**.
	e.g., Wire a numeric control **input** to a numeric indicator **output**.
	e.g., LabVIEW has its own directory with extension **.llb**.
	e.g., The function **Build Array** can build an array.
	e.g., The **Case** structure is in the subpalette **Structures**.
	e.g., The while loop in C++ is different from the **While Loop** in LabVIEW.
	e.g., Choose **Save** from the pull down menu **File**.
	e.g., Choose **Latch When Released** as the mechanical action.
	e.g., The vector **x** is the first column of the matrix **X**.

Guidelines for the Readers XV

italic	emphasized text e.g., You *must* include the folder **Shared Libraries** in the same folder where the executable is saved.
`Courier`	The text you enter from the keyboard. e.g., Enter `Action List` in **Item Name**. Note that if the text you entered becomes an item, it will be in **bold** font. For example, if you enter a text `Option 1` in a **Text Ring**, then it will become an item **Option 1**.
>>	Path of files and LabVIEW functions on all platforms. e.g., Find **Functions** >> **Numeric** >> **Add**. e.g., Find the example, LabVIEW directory >> **examples** >> **general** >> **arrays.llb** >> **Building Arrays.vi**.
Ctrl-H	Key combination of Control key and H. The Control key can be replaced by the Command key for Macintosh platforms. e.g., Press **Ctrl-H**, which is equivalent to **Command-H** on Macintosh platforms, for online help.

Acronyms

B	Battery e.g., Pin 1: B V+, Pin 2: B V– means connect Pin 1 with positive of the battery, and Pin 2 with negative of the battery.
FG	Function Generator e.g., Pin 43: FG V+, Pin 50: FG V– means connect Pin 43 with positive lead of Function Generator, and Pin 50 with negative lead of Function Generator.
Hz	Hertz
ms	millisecond

Issues on Macintosh Platforms

As for the key combination, the Control key can be replaced by the Command key for Macintosh platforms. Right clicking on the mouse is equivalent to **Command**-clicking on Macintosh platforms. Each directory in a path is conventionally separated by a backward slash (\) except for Macintosh platforms, where a colon (:) is used. For example, **c:\my_folder\new_VIs** would be equivalent to **c:my_folder:new_VIs** on Macintoshes. As for VI compatibility, you can transfer VIs across different platforms at your will. However, some functionality may not be applicable to different platforms; for example, if your VI contains sub VIs for Active X, it will not function correctly on Macintosh platforms since Macintosh platforms do not support Active X. Otherwise, the compatibility of VIs is transparent across different platforms, including Macintoshes.

Compatible Versions of LabVIEW

All of the example VIs are written in LabVIEW 5.0; therefore, LabVIEW 5.x or higher will be able to *open* them. However, all of the techniques and VIs can be realized in LabVIEW 3.x or higher. Also, since all of the examples are kept simple, you can easily duplicate them because they are written using standard LabVIEW VIs and functions. As for the data acquisition examples, they are the modified versions of LabVIEW examples. Therefore, you can easily duplicate them, too, by following the step-by-step instructions provided in this manual.

As the newer version appears, some of the names or paths (location of VIs) may not match. For example, the subpalette **Functions** >> **Analysis** has been divided into two new subpalettes **Signal Processing** and **Mathematics** with some new VIs in LabVIEW 5.1. Therefore, you should look for the two new subpalettes to find the analysis VIs if you are using LabVIEW 5.1 or higher.

Regardless of the version of your LabVIEW, however, this exercise manual is written in the most general way so that any difference in different versions should have no effect on using the drill problems except for the minor VI paths or names. This is due to the philosophy behind this manual and the main text: simplicity with rich applicability. Most of the differences in different versions of LabVIEW are minor, and you can easily catch up with them once you master this manual as well as the main text. Therefore, the differences in different LabVIEW versions will have no effect on both the main text and this exercise manual except for some minor VI paths or names. The information about such differences can be found in your LabVIEW package.

Hardware Configuration

In all drill problems about data acquisition and instrument control, it is assumed that your data acquisition board and GPIB board are properly configured regardless of the type and the vendor. If you are using LabPC-1200, configure the board with the following settings in NI-DAQ configuration utility: 1) analog input mode as Differential and 2) analog output Mode as Bipolar.

Screen Shots of VIs

All of the screen shots of VIs in this manual are used with the permission of National Instruments.

Drill Problems
Day 1

Drill Problem 1.1

1. VIs to be used: **p021_WirePrac.vi** (Template provided.)
2. Objective: To learn the correct and accurate wiring technique. The correct wiring technique can eliminate most of the programming errors.
3. Estimated time: 10–15 minutes
4. Related chapter: Chapter 2
5. Key objects, VIs, and functions in this drill problem:
   Functions >> Data Acquisition >> Analog Input >> AI Config.vi and **AI Start.vi**
6. Instructions

   Open the VI and create wires, controls, and indicators as shown in Figure D1.1. Feel free to try other VIs and practice wiring on them. An easy way to create controls or indicators for each terminal is to right click (**Command**-click on Macintosh platforms) on the terminal to pop up the menu and select **Create Control** or **Create Indicator**.

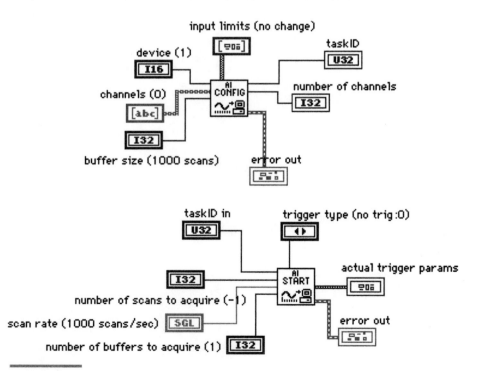

Figure D1.1

Drill Problem 1.2

1. VIs to be used: **p022_WirePrac.vi** (Template provided.)
2. Objective: Continuation of wiring practice. Try to get used to complicated diagrams. Learn to feel comfortable with complicated connections and many sub VIs by spending some time on examining such diagrams. Observe the **Run** button as you wire terminals. If it is broken, clicking the button will bring up an error list. Double clicking on each list will take you to the location where that error is occurred.
3. Estimated time: 10–15 minutes
4. Related chapter: Chapter 2
5. Key objects, VIs, and functions in this drill problem:
   Functions >> Data Acquisition >> Analog Input >> AI Config.vi, AI Start.vi, AI Read.vi, and **AI Clear.vi**
   Functions >> Time & Dialog >> General Error Handler.vi
6. Instructions
   Open the VI and create wires, controls, and indicators as shown in Figure D1.2. Use the pop-up menu at each terminal to create controls and indicators.

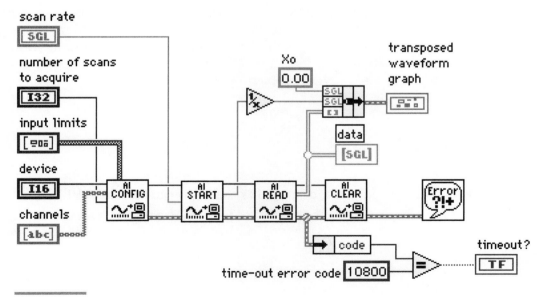

Figure D1.2

Drill Problem 1.3

1. VIs to be used: **p023_ShrtCutPrac.vi** (Template provided.)
2. Objective: To practice the shortcuts described in Chapter 2. Try to use them at all times, since they can prevent you from making minor to critical mistakes in programming with LabVIEW. Also, try to remember where you found the VIs in this example.
3. Estimated time: 10–15 minutes
4. Related chapter: Chapter 2
5. Key objects, VIs, and functions in this drill problem:

    Functions >> **Data Acquisition** >> **Analog Input** >> **AI Start.vi**

    Functions >> **File I/O** >> **Open/Create/Replace File.vi** and **Write To Spreadsheet File.vi**

    Functions >> **Data Acquisition** >> **Digital I/O** >> **Write to Digital Port.vi**

    Functions >> **Analysis** >> **Measurement** >> **Impulse Response Function.vi**

    Functions >> **Data Acquisition** >> **Counter** >> **Counter Events or Time.vi**

6. Instructions

    Open the VI and double click on each sub VI to open its front panel. Using the shortcuts, navigate between the front panels and the diagram windows of different sub VI levels. Continue to practice until you feel comfortable with the shortcuts. They will reduce a great deal of developing time and prevent problematic situations.

Drill Problem 1.4

1. VIs to be used: **p031_dBcalc.vi** (You need to create a new VI.)
2. Objective: To create the first VI of your own. Try to use all the shortcuts and the technique of creating controls and indicators from the pop-up menu of each terminal.
3. Estimated time: 15–20 minutes
4. Related chapter: Chapter 3

Drill Problems • Day 1   5

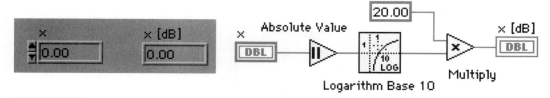

Figure D1.4

5. Key objects, VIs, and functions in this drill problem:
   Functions >> Numeric >> Absolute Value and **Multiply**
   Functions >> Numeric >> Logarithmic >> Logarithm Base 10
6. Instructions
   Complete the front panel and the diagram window of the VI as shown in Figure D1.4. Lay out the wire connections as clearly as possible. In other words, make it obvious which wire is connected to which terminal. Save the VI, and it will be used as a sub VI in Drill Problem 1.6. Refer to Chapter 3 to learn how to create terminals for each control and indicator.

Drill Problem 1.5

1. VIs to be used: **p032_QuitPrompt.vi** (You need to create a new VI.)
2. Objective: To learn more about the options in the **VI Setup . . .** window. The options selected in this drill problem are commonly used in many applications.
3. Estimated time: 25–30 minutes
4. Related chapter: Chapter 3
5. Key objects, VIs, and functions in this drill problem:
   Functions >> Structures >> While Loop
6. Instructions
   Step 1: Open a new VI by choosing **New** from the pull-down menu **File**.
   Step 2: Place a **String Control** from **Controls >> String & Table** in the front panel, and enter Quitting . . . Press Yes to Complete. Adjust the size of font and justification from **Text Settings** menu ring ⎡9pt Application Font ▼⎤. First, select the text in the **String Con-**

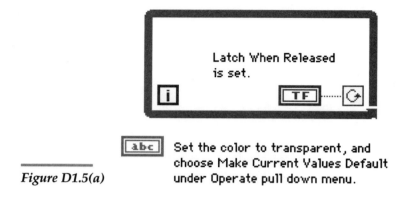

Figure D1.5(a)

Set the color to transparent, and choose Make Current Values Default under Operate pull down menu.

trol using the Text Editor . Then use **Size** and **Justify** under the **Text Settings** menu ring to change the font size and justification.

Step 3: Go to the diagram window, and complete the VI as shown in Figure D1.5(a).

First, place a **While Loop** from **Function >> Structures.**

Second, right click on the **conditional terminal** , and select **Create Control** from its pop-up menu to create a Boolean switch.

Third, go to the front panel, and locate the Boolean switch you just created. When you find it, right click on it, and set it to **Latch When Released** by selecting **Mechanical Action >> Latch When Released** from its pop-up menu.

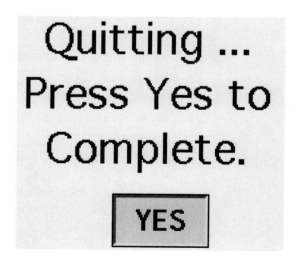

Figure D1.5(b)

Drill Problems • Day 1

Fourth, press the Boolean switch to set it to the state TRUE.

Fifth, choose **Make Current Values Default** under the pull-down menu **Operate**, or **Data Operations** >> **Make Current Value Default** from the pop-up menu of the Boolean switch. The second method is how you change the default value of an individual object.

It will be assumed that those five steps have always been performed on the Boolean switch every time you use a **While Loop**. In this drill problem, change the text of the Boolean switch to YES for both TRUE and FALSE states.

Step 4: Adjust the size of the front panel to show only the text and the Boolean switch as shown in Figure D1.5(b).

Step 5: In the front panel, select **VI Setup . . .** from the pop-up menu as shown in Figure D1.5(c).

Step 6: Under **Execution Options** in the **VI Setup** window, check only **Show Front Panel When Called** and **Close Afterwards if Originally Closed**.

Step 7: Under **Window Options**, check only **Auto-Center**. Then click OK to exit the window.

Step 8: Save the VI as **p032_QuitPrompt.vi**. This VI will be used in the next drill problem. You do not need to create any terminal for this sub VI.

Figure D1.5(c)

Drill Problem 1.6

1. VIs to be used: **p033_RNDisplay.vi** (Template provided.)
2. Objective: To build a VI that displays random numbers on a chart in real time, using the two sub VIs built in Drill Problems 1.5 and 1.6.

3. Estimated time: 25–30 minutes
4. Related chapter: Chapter 3
5. Key objects, VIs, and functions in this drill problem:
   Controls >> Graph >> Waveform Chart
   Functions >> Time & Dialog >> Wait (ms)
   Functions >> Numeric >> Random Number (0-1) and **Quotient & Remainder**
   Functions >> Comparison >> Equal To 0?
   Functions >> Structures >> While Loop and **Case**
6. Instructions

   Complete both the front panel and the diagram window as shown in Figure D1.6. There is an Attribute Node of **Waveform Chart** in use. You can create it by selecting **Create >> Attribute Node** from its pop-up menu, but it is already included in the template VI for you. The other frame of each **Case** structure is empty. Make sure to have the input of the second **Case** wired to the output tunnel to complete the **Case** structure. (You must specify the outputs in both frames of the **Case** structures.) A detailed discussion of the **Case** structure is presented in Chapter 4.

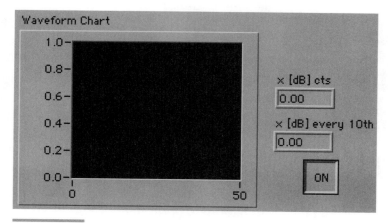

Figure D1.6(a)

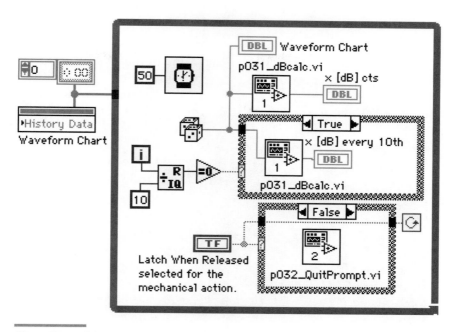

Figure D1.6(b)

Drill Problem 1.7

1. VIs to be used: **p041_ForLp&Indxng.vi** (Template provided.)
2. Objective: To study the effect of *indexing* in LabVIEW. Indexing is useful when you need to create an array from a stream of data samples or when you need each individual sample in a data array.
3. Estimated time: 25–30 minutes
4. Related chapter: Chapter 4
5. Key objects, VIs, and functions in this drill problem:
   Controls >> Graph >> XY Graph
   Functions >> Structures >> For Loop
   Functions >> Numeric >> Random Number (0-1)
   Functions >> Time & Dialog >> Wait (ms)
   Functions >> Cluster >> Bundle

6. Instructions

    Complete the VI as shown in Figure D1.7. Observe the effect of indexing at each boundary of the two **For Loop**s. You may turn the auto indexing feature on or off by right clicking on the black dot, which is called *tunnel*, on the boundary of the **For Loop**, and selecting either **Enable Indexing** or **Disable Indexing**, respectively.

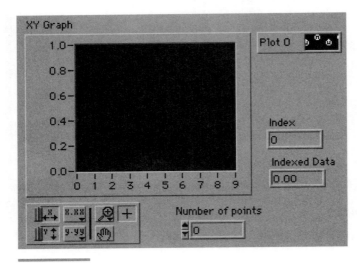

Figure D1.7(a)

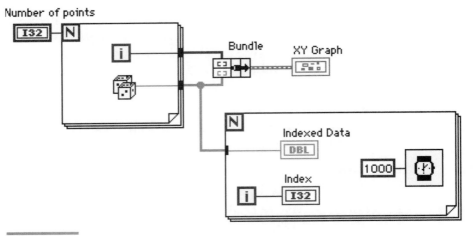

Figure D1.7(b)

Drill Problem 1.8

1. VIs to be used: **p042_WhileLp&ShftReg.vi** (Template provided.)
2. Objective: To study the following three topics: (1) shift register, (2) indexing arrays, and (3) multichannel display on **Waveform Chart**s using **Bundle**. This drill problem will show that the average of random Gaussian numbers approaches their true mean as you have larger number of samples in computation. Also, the running average is used. It returns the latest average value at each new data sample instead of collecting the entire samples to compute the mean. The equation for the running average is given in Chapter 14.
3. Estimated time: 30–35 minutes
4. Related chapter: Chapter 4
5. Key objects, VIs, and functions in this drill problem:
   Controls >> Graph >> Waveform Chart
   Functions >> Structures >> While Loop
   Functions >> Analysis >> Signal Generation >> Gaussian White Noise.vi
   Functions >> Array >> Index Array
   Functions >> Time & Dialog >> Wait (ms)
   Functions >> Cluster >> Bundle
6. Instructions

   Complete the diagram window as shown in Figure D1.8. Study the role of the shift register, which can be created by right clicking on the boundary of the **While Loop** and selecting **Add Shift Register**. That will create a pair on both sides. There is only one element on the right side (the one with the up arrow), but you can have multiple elements on the left side by dragging down the one with the down arrow. The purpose of the Attribute Node of **chart** is to refresh the display before starting the **While Loop**.

   In the front panel, set the mechanical action of the Boolean switch **Quit** to **Latch When Released**. Make sure to set the switch to TRUE state, and choose **Make Current Values Default** under the pull-down menu **Operate** to change its default state. This actually sets the current values of all the objects in the front panel as their default value. To change the default value of an individual object, right click on the one of interest, and choose **Data Operations >> Make Current Value Default** from its pop-up menu. That will change the default value of only the object you right clicked on.

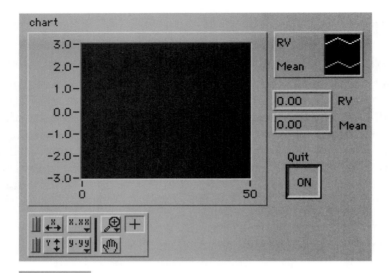

Figure D1.8(a)

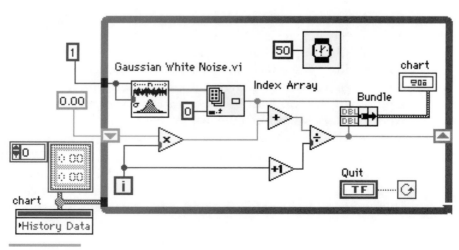

Figure D1.8(b)

Drill Problem 1.9

1. VIs to be used: **p043_Loops&Conditions.vi** (Template provided.)
2. Objective: To build a more complicated VI. You will acquire a single data sample at every 300 msec while performing moving average (MA) of window size 3 and displaying both the averaged and the original data. The signal level is also monitored and indicated on an LED. This drill problem also shows how to display multiple channels of data on a **Waveform Chart**.
3. Estimated time: 30–35 minutes
4. Related chapter: Chapter 4
5. Key objects, VIs, and functions in this drill problem:
   Controls >> Graph >> Waveform Chart
   Functions >> Structures >> While Loop
   Functions >> Numeric >> Random Number (0-1), **Divide**, and **Compound Arithmetic**

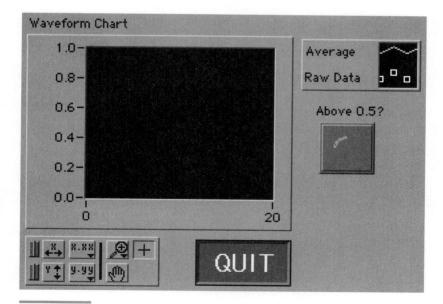

Figure D1.9(a)

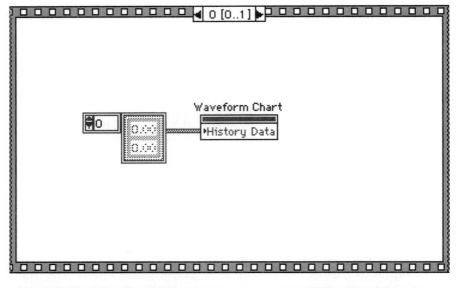
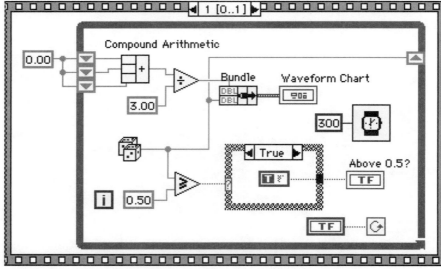
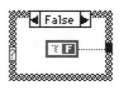

Figure D1.9 (b)

Functions >> Comparison >> Greater Or Equal?
Functions >> Time & Dialog >> Wait (ms)
Functions >> Cluster >> Bundle

6. Instructions:

   Open the template VI, and complete it as shown in Figure D1.9. The frame 0 of the **Sequence** structure contains an Attribute Node of **Waveform Chart**. The False frame of the **Case** structure has the FALSE Boolean constant wired to **Above 0.5?**.

Drill Problems
Day 2

Drill Problem 2.1

1. VIs to be used: **s051_MechActionOfBooleans.vi** (The VI is already complete.)
2. Objective: To examine the different types of mechanical action settings of Boolean switches.
3. Estimated time: 10–15 minutes
4. Related chapter: Chapter 5
5. Key objects, VIs, and functions in this drill problem:
   N/A
6. Instructions
   Open the VI and observe the effect of different mechanical action settings. The VI is complete already.

Drill Problem 2.2

1. VIs to be used: **p052_MultiChChartGraph.vi** (Template provided.)
2. Objective: To learn how to plot multichannels of data on a **Waveform Chart** and a **Waveform Graph.**
3. Estimated time: 20–25 minutes
4. Related chapter: Chapter 5
5. Key objects, VIs, and functions in this drill problem:
   Controls >> Graph >> Waveform Chart and **Waveform Graph**

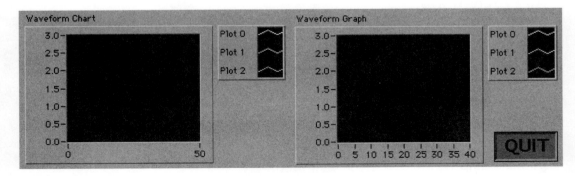

Figure D2.2(a)

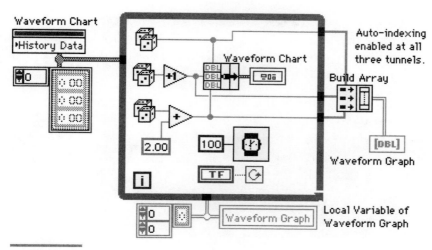

Figure D2.2(b)

      Functions >> Array >> Build Array
      Functions >> Cluster >> Bundle

6. Instructions

   Complete the VI as shown in Figure D2.2. There are a **Waveform Chart** and a **Waveform Graph**, inside and outside the **While Loop**, respectively. To create a constant for the Attribute Node of **Waveform Chart** and the **Local Variable** of **Waveform Graph**, use the pop-up menu of each.

Drill Problem 2.3

1. VIs to be used: **p053_MultiChGphXo.vi** (Template provided.)
2. Objective: To learn how to plot multichannels of data on a **Waveform Graph** with different starting points and intervals of x-axis.
3. Estimated time: 30–35 minutes
4. Related chapter: Chapter 5
5. Key objects, VIs, and functions in this drill problem:
   Functions >> Cluster >> Bundle
   Functions >> Array >> Build Array

6. Instructions

   Complete the VI as shown in Figure D2.3. The Boolean switch wired to the *conditional terminal* of the **While Loop** has the default value of TRUE with mechanical action **Latch When Released**.

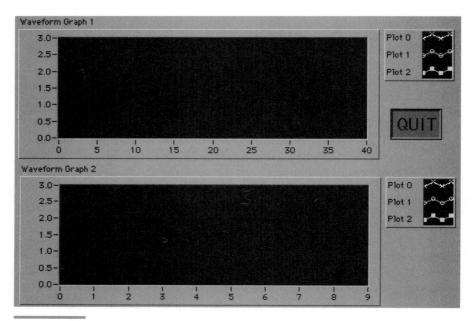

Figure D2.3(a)

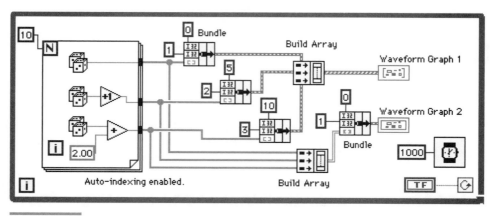

Figure D2.3(b)

Drill Problems • Day 2

Drill Problem 2.4

1. VIs to be used: **p054_MultiChXYGph.vi** (Template provided.)
2. Objective: To learn how to plot multichannels of data on an **XY Graph** and compare its functionality with that of **Waveform Graph**.
3. Estimated time: 30–35 minutes
4. Related chapter: Chapter 5
5. Key objects, VIs, and functions in this drill problem:
   Controls >> Graph >> XY Graph and **Waveform Graph**
   Functions >> Cluster >> Bundle
   Functions >> Array >> Build Array
6. Instructions
   Complete the VI as shown in Figure D2.4. Note that the Boolean switch in the **While Loop** has the default value of TRUE with mechanical action **Latch When Released**.

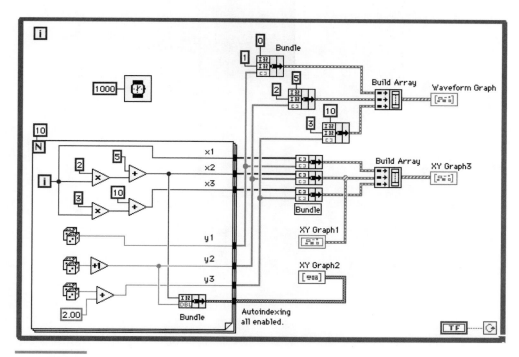

Figure D2.4(a)

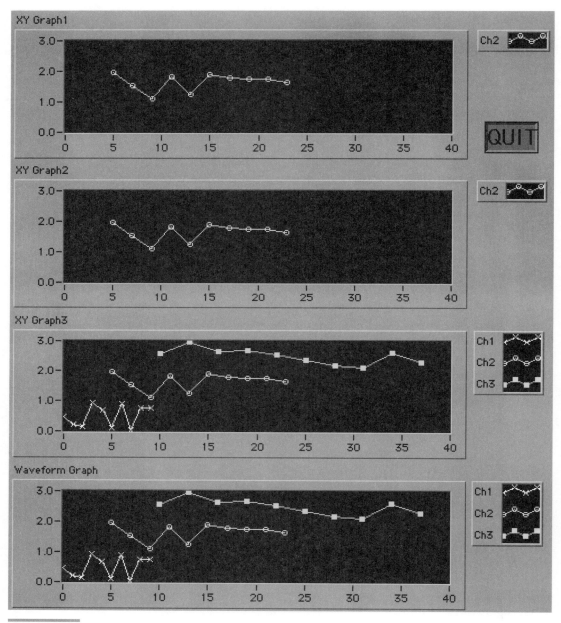

Figure D2.4(b)

Drill Problem 2.5

1. VIs to be used: **p061_ArryPrac.vi** (You need to create a new VI.)
2. Objective: To learn how to build a matrix or a vector from multiple vectors and study the different input settings of **Build Array**.
3. Estimated time: 15–20 minutes
4. Related chapter: Chapter 6
5. Key objects, VIs, and functions in this drill problem:
   Controls >> **Array & Cluster** >> **Array**
   Controls >> **Numeric** >> **Digital Control** and **Digital Indicator**
   Functions >> **Array** >> **Build Array**

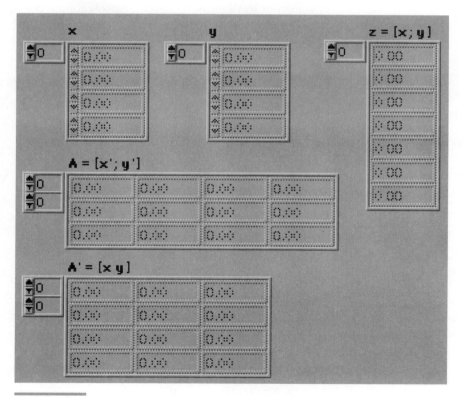

Figure D2.5(a)

24 Drill Problems • Day 2

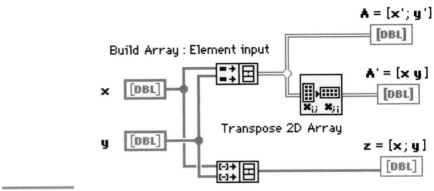

Figure D2.5(b)

6. Instructions
   Complete the VI as shown in Figure D2.5. Run the VI to see the effect of two different input settings of **Build Array:** Element Input and Array Input. Element Input setting increases the dimension of the input array, whereas Array Input setting concatenates the input arrays, maintaining the same dimension.

Drill Problem 2.6

1. VIs to be used: **p062_ClustrFcnPrac.vi** (You need to create a new VI.)
2. Objective: To learn how to build a cluster of arrays and extract arrays from it conversely. You may skip this drill problem if necessary.
3. Estimated time: 15–20 minutes
4. Related chapter: Chapter 6
5. Key objects, VIs, and functions in this drill problem:
   Controls >> **Array & Cluster** >> **Cluster** and **Array**
   Functions >> **Cluster** >> **Build Cluster Array** and **Index & Bundle Cluster Array**
6. Instructions
   Complete the VI as shown in Figure D2.6. "cluster(x)" indicates a cluster whose element is an array x. "cluster(1,4,7)" indicates a cluster whose elements are the 1st, 4th, and 7th elements of the input arrays.

The order of each element is as follows: **x** has the 1st, 2nd, and 3rd elements; **y** has the 4th, 5th, and 6th elements; and **z** has the 7th, 8th, and 9th elements. In this drill problem, the order of each element happens to be identical to the value itself.

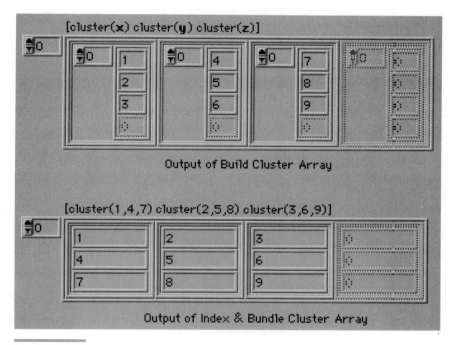

Figure D2.6(a)

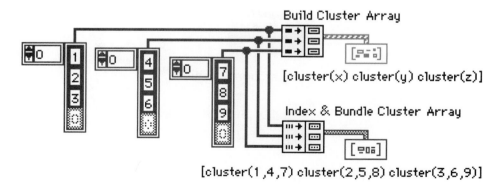

Figure D2.6(b)

Drill Problem 2.7

1. VIs to be used: **p063_IntstyGphPxlByPxl.vi** (Template provided.)
2. Objective: To learn how to display 2-D array data using an **Intensity Graph**, which can especially be useful if the 2-D array data represent an image. You may skip this drill problem if necessary.
3. Estimated time: 40 minutes
4. Related chapter: Chapter 6
5. Key objects, VIs, and functions in this drill problem:
   Controls >> Graph >> Intensity Graph
   Controls >> Numeric >> Color Box

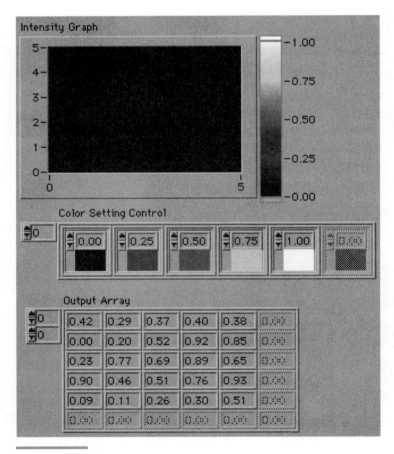

Figure D2.7(a)

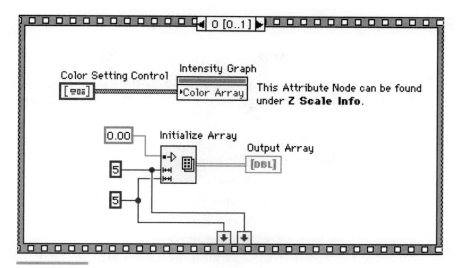

Figure D2.7(b)

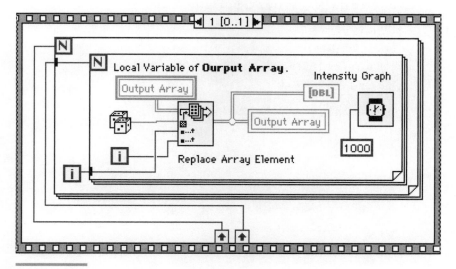

Figure D2.7(c)

    Controls >> Array & Cluster >> Array
    Functions >> Structures >> Sequence
    Functions >> Array >> Initialize Array and **Replace Array Element**

6. Instructions

    Complete the VI as shown in Figure D2.7.

Drill Problem 2.8

1. VIs to be used: **p064_IntstyGphAttNode.vi** (Template provided.)
2. Objective: To learn more about how to display 2-D array data using an **Intensity Graph**. You may skip this drill problem if necessary.
3. Estimated time: 40 minutes
4. Related chapter: Chapter 6
5. Key objects, VIs, and functions in this drill problem:
   Controls >> **Graph** >> **Intensity Graph**
   Controls >> **Numeric** >> **Color Box**
   Controls >> **Array & Cluster** >> **Array**
   Functions >> **Structures** >> **Sequence**
   Functions >> **Numeric** >> **Trigonometric** >> **Sine** and **Cosine**
6. Instructions
   Complete the VI as shown in Figure D2.8.

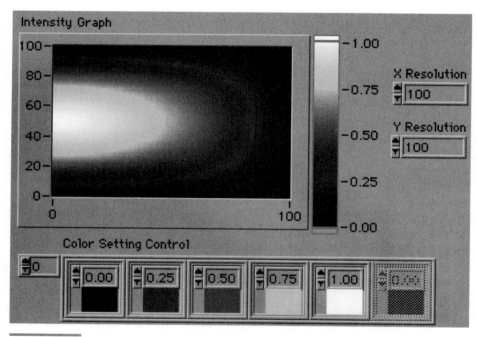

Figure D2.8(a)

Drill Problems • Day 2

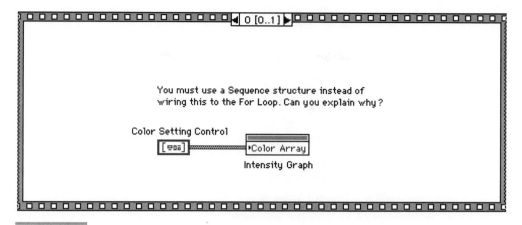

Figure D2.8(b)

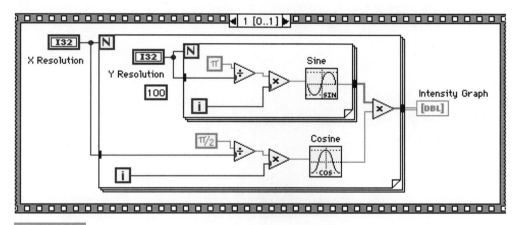

Figure D2.8(c)

Drill Problem 2.9

1. VIs to be used: **p065_Arry&ClustrUpdPrac.vi** (Template provided.)
2. Objective: To learn how to index elements in clusters and arrays.
3. Estimated time: 40 minutes
4. Related chapter: Chapter 6
5. Key objects, VIs, and functions in this drill problem:
   Controls >> Numeric >> Digital Control

Controls >> Boolean >> Square LED and **Round LED**
Controls >> Array & Cluster >> Array
Functions >> Cluster >> Unbundle and **Unbundle By Name**
Functions >> Array >> Replace Array Element

6. Instructions

   Complete the VI as shown in Figure D2.9. This drill problem consists of two parts: updating elements in a cluster, and indexing elements in an array. The first part is the left portion of the front panel, and the second part, the right portion of the front panel.

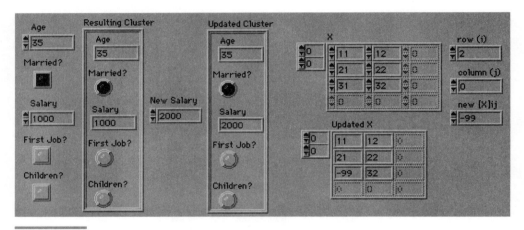

Figure D2.9(a)

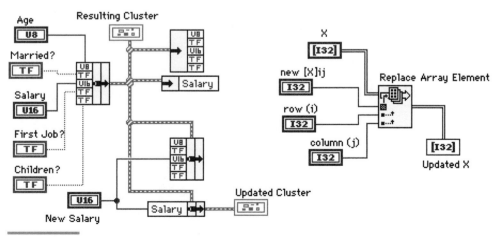

Figure D2.9(b)

Drill Problems
Day 3

Drill Problem 3.1

1. VIs to be used: **s081_AcqNScans.vi** (The VI is already complete.)
2. Objective: To learn how to use built-in examples of LabVIEW for one snapshot of analog input acquisition.
3. Estimated time: 10–15 minutes
4. Related chapter: Chapter 8
5. Key objects, VIs, and functions in this drill problem:
   VIs in **Functions >> Data Acquisition >> Analog Input**
6. Instructions

   Open the VI and examine its sub VIs, and different parameters for analog input data acquisition. Read the information under **Windows >> Show VI Info . . . .** (You can also use the shortcut **Ctrl-I** to bring up the information.) Follow the comments under NOTE to correctly address input channels. The input channel(s) must be listed in reverse order and channel 0 must be included for multichannel scanning using LabPC-1200.

   For more information about the VI, refer to the example **Acquire N Scans.vi**, which can be found in LabVIEW directory >> **examples** >> **daq** >> **anlogin** >> **anlogin.llb**.

Drill Problem 3.2

1. VIs to be used: **s082_CtsAcq&Chart(buff).vi** (The VI is already complete.)
2. Objective: To learn how to use built-in examples of LabVIEW for buffered continuous analog input data acquisition.
3. Estimated time: 10–15 minutes
4. Related chapter: Chapter 8
5. Key objects, VIs, and functions in this drill problem:
   VIs in **Functions >> Data Acquisition >> Analog Input**

6. Instructions

   Open the VI and examine its sub VIs, and different parameters for buffered continuous analog input data acquisition. Read the information under **Windows >> Show VI Info . . . .** (You can also use the shortcut **Ctrl-I** to bring up the information.) Follow the comments under NOTE to correctly address input channels. The input channel(s) must be listed in reverse order and channel 0 must be included for multichannel scanning using LabPC-1200.

   For more information about the VI, refer to the example **Cont Acq&Chart (buffered).vi**, which can be found in LabVIEW directory >> examples >> daq >> anlogin >> anlogin.llb.

Drill Problem 3.3

1. VIs to be used: **s091_Gen1PtOn1Chwith_s082.vi** (The VI is already complete.)
2. Objective: To learn how to use built-in examples of LabVIEW for DC generation (1 point analog output). The VI **s082_CtsAcq&Chart (buff).vi** is used to monitor the generated dc signal.
3. Estimated time: 10–15 minutes
4. Related chapter: Chapter 9
5. Key objects, VIs, and functions in this drill problem:
   VIs in **Functions >> Data Acquisition >> Analog Output**
6. Instructions

   Open the VI and examine its sub VIs. Read the information under **Windows >> Show VI Info . . .** for signal and pin connections. (You can also use the shortcut **Ctrl-I** to bring up the information.) This drill problem uses **s082_CtsAcq&Chart(buff).vi** as well. Therefore, open **s082_CtsAcq&Chart(buff).vi**, and run it together with this drill problem.

   For more information about the VI, refer to the example **Generate 1 Point on 1 Channel.vi**, which can be found in LabVIEW directory >> examples >> daq >> anlogout >> anlogout.llb.

Drill Problem 3.4

1. VIs to be used: **s092_FGwith_s082.vi** (The VI is already complete.)
2. Objective: To learn how to use built-in examples of LabVIEW for waveform generation. The VI **s082_CtsAcq&Chart(buff).vi** is used to monitor the generated waveforms.
3. Estimated time: 10–15 minutes
4. Related chapter: Chapter 9
5. Key objects, VIs, and functions in this drill problem:
   VIs in **Functions >> Data Acquisition >> Analog Output**
6. Instructions
   Open the VI and examine its sub VIs. Read the information under **Windows >> Show VI Info ...** for signal and pin connections. (You can also use the shortcut **Ctrl-I** to bring up the information.) This drill problem uses **s082_CtsAcq&Chart(buff).vi** as well. Therefore, open **s082_CtsAcq&Chart(buff).vi**, and run it together with this drill problem.
   For more information about the VI, refer to the example **Function Generator.vi**, which can be found in LabVIEW directory **>> examples >> daq >> anlogout >> anlogout.llb**.

Drill Problem 3.5

1. VIs to be used: **s101_CtsPTrn(8253)with_s082.vi** (The VI is already complete.)
2. Objective: To learn how to use built-in examples of LabVIEW for pulse train generation. The VI **s082_CtsAcq&Chart(buff).vi** is used to monitor the generated pulse train.
3. Estimated time: 10–15 minutes
4. Related chapter: Chapter 10
5. Key objects, VIs, and functions in this drill problem:
   VIs in **Functions >> Data Acquisition >> Counter**

6. Instructions

    Open the VI and examine its sub VIs. Read the information under **Windows >> Show VI Info...** for signal and pin connections. (You can also use the shortcut **Ctrl-I** to bring up the information.) This drill problem uses **s082_CtsAcq&Chart(buff).vi** as well. Therefore, open **s082_CtsAcq&Chart(buff).vi**, and run it together with this drill problem.

    For more information about the VI, refer to the example **Cont Pulse Train (8253).vi**, which can be found in LabVIEW directory >> **examples >> daq >> counter >> 8253.llb**.

Drill Problem 3.6

1. VIs to be used: **s102_CntEvnts(8253).vi** (The VI is already complete.)
2. Objective: To learn how to use built-in examples of LabVIEW for event counting.
3. Estimated time: 10–15 minutes
4. Related chapter: Chapter 10
5. Key objects, VIs, and functions in this drill problem:

    VIs in **Functions >> Data Acquisition >> Counter**
6. Instructions

    Open the VI and examine its sub VIs. Read the information under **Windows >> Show VI Info...** for signal and pin connections. (You can also use the shortcut **Ctrl-I** to bring up the information.)

    For more information about the VI, refer to the example **Count Events (8253).vi**, which can be found in LabVIEW directory >> **examples >> daq >> counter >> 8253.llb**.

Drill Problem 3.7

1. VIs to be used: **p111_WrtRdBin1D.vi** (Template provided.)
2. Objective: To learn how to save and read a 1-D binary data file.
3. Estimated time: 30–35 minutes

4. Related chapter: Chapter 11
5. Key objects, VIs, and functions in this drill problem:
   VIs in **Functions >> File I/O**
   Functions >> Time & Dialog >> Simple Error Handler.vi
6. Instructions
   Complete the VI as shown in Figure D3.7. Frame 0 saves 1-D data to a file in binary format and frame 1 reads them back and displays them. View the data file using a text editor such as Notepad to see if the data are understandable.

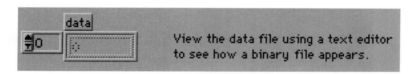

Figure D3.7(a)

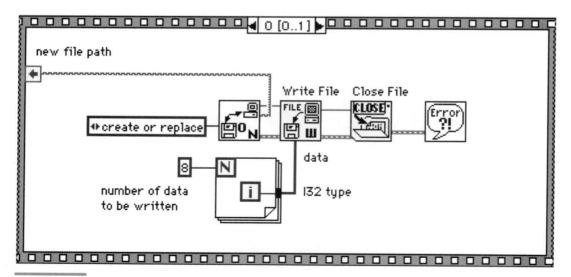

Figure D3.7(b)

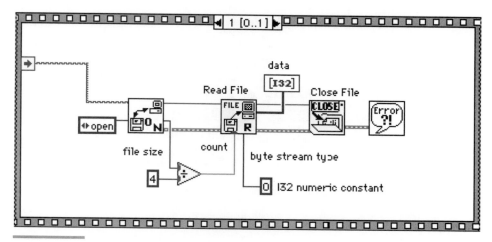

Figure D3.7(c)

Drill Problem 3.8

1. VIs to be used: **p112_WrtRdASCII.vi** (Template provided.)
2. Objective: To learn how to save and read an ASCII data file.
3. Estimated time: 30–35 minutes
4. Related chapter: Chapter 11
5. Key objects, VIs, and functions in this drill problem:
   VIs in **Functions >> File I/O**
   Functions >> Time & Dialog >> Simple Error Handler.vi
6. Instructions

   Complete the VI as shown in Figure D3.8. Frame 0 saves the string constant **Testing Message** in ASCII format and frame 1 reads it back and displays it. View the data file using a text editor such as Notepad to see if the data are understandable.

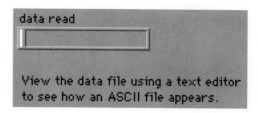

Figure D3.8(a)

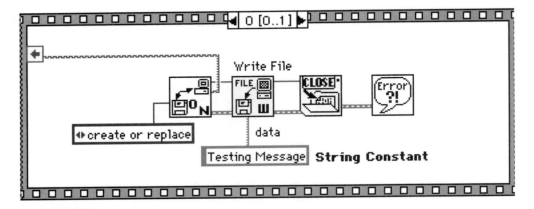

Figure D3.8(b)

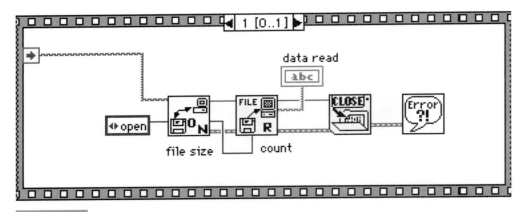

Figure D3.8(c)

Drill Problem 3.9

1. VIs to be used: **p121_TypeCast.vi** (Template provided.)
2. Objective: To learn how to convert binary string data to numeric data.
3. Estimated time: 30–35 minutes
4. Related chapter: Chapter 12
5. Key objects, VIs, and functions in this drill problem:
   Controls >> List & Ring >> Text Ring

Drill Problems • Day 3

Controls >> String & Table >> String Indicator
VIs in **Functions** >> **File I/O**
Functions >> **Advanced** >> **Data Manipulation** >> **Type Cast**

6. Instructions

    Complete the VI as shown in Figure D3.9. Frame 0 writes a binary data file, frame 1 reads it back as string data (binary string data), and frame 2 converts the data to numeric data.

    To get the text ring **Data type** in the front panel, follow these steps:

    Step 1: Place a **Text Ring** from **Controls** >> **List & Ring** in the front panel.

    Step 2: Using the Text Editor ▣, enter UB (1 byte) followed by **Shift**-Return or **Shift**-Enter.

    Step 3: Enter the next item I32 (4 byte) followed by **Shift**-Return or **Shift**-Enter.

    Step 3: Enter the last item DBL (8 byte).

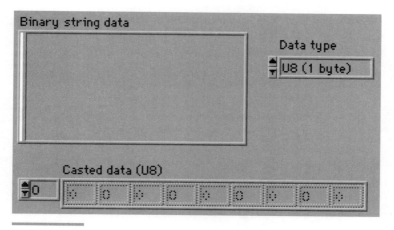

Figure D3.9(a)

There are two **Case** structures (each in frame 0 and 2 of the **Sequence**), and frame 1 of each **Case** structure is chosen as default. To change the default frame from 0 to 1, go to frame 1, right click on the frame label, and select **Make This The Default Case** from its pop-up menu.

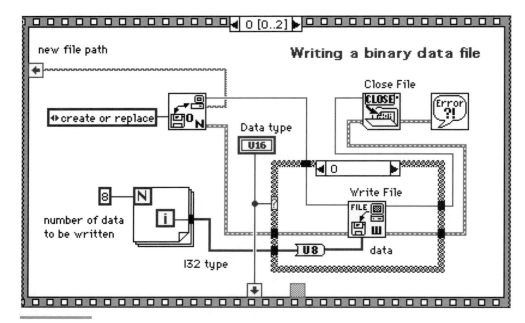

Figure D3.9(b)

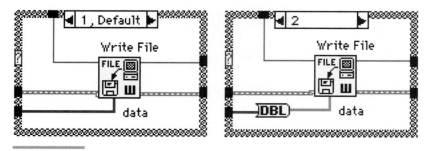

Figure D3.9(c)

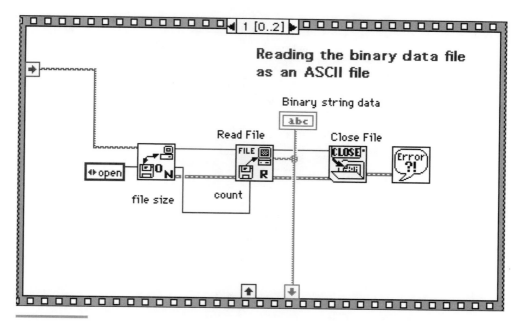

Figure D3.9(d)

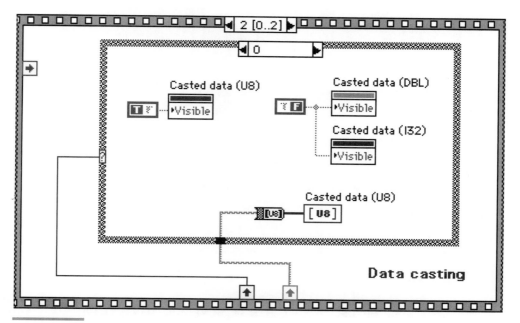

Figure D3.9(e)

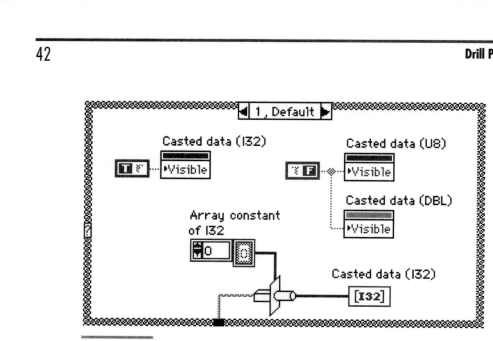

Figure D3.9(f)

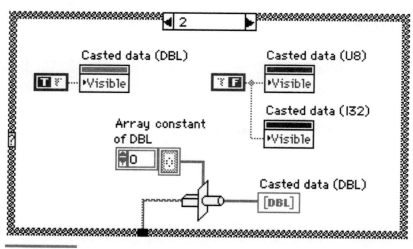

Figure D3.9(g)

Drill Problem 3.10

1. VIs to be used: **p141_AliasTest.vi** and **p141_FFTproc.vi** (Templates provided.)

Drill Problems • Day 3

2. Objective: To learn how to modify LabVIEW examples for your application. This drill problem adds frequency analysis capability to the VI **s082_CtsAcq&Chart(buff).vi**.
3. Estimated time: 30–35 minutes
4. Related chapter: Chapter 14
5. Key objects, VIs, and functions in this drill problem:
   Functions >> Analysis >> Digital Signal Processing >> Real FFT.vi
   Functions >> Analysis >> Array Operation >> Quick Scale 1D.vi
6. Instructions

   Open **p141_AliasTest.vi**, and complete the **While Loop** as shown in Figure D3.10(a) and (b).

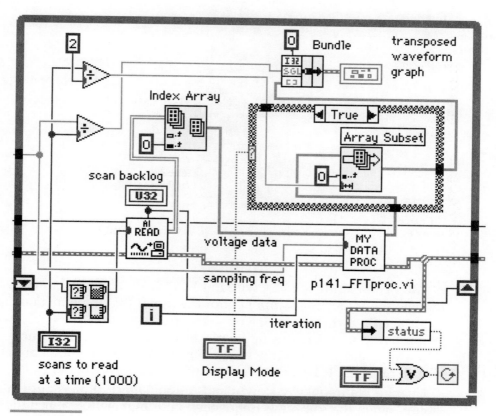

Figure D3.10(a)

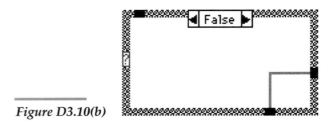

Figure D3.10(b)

Open **p141_FFTproc.vi** and complete it as shown in Figure D3.10(c)–(e). Note that two new terminals **iteration** and **sampling freq: fs** need to be created.

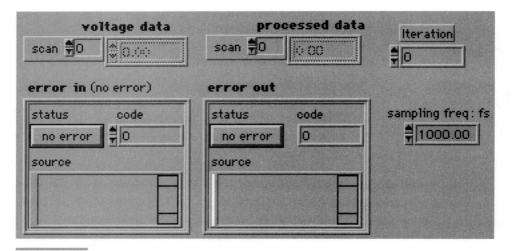

Figure D3.10(c)

After completing the VIs, wire Pin 1 to FG V+ and Pin 2 to FG V−, and view the frequency component of the signal that the Function Generator sends to the board. If you are not using LabPC-1200, connect the positive pin and the negative pin of any analog input channel to the positive and the negative of the Function Generator to complete this drill problem. FG is an abbreviation for Function Generator.

Change the frequency slowly and see the effect of aliasing, which occurs when the sampling frequency is not higher than twice the maximum frequency (frequency bandwidth) of the input signal.

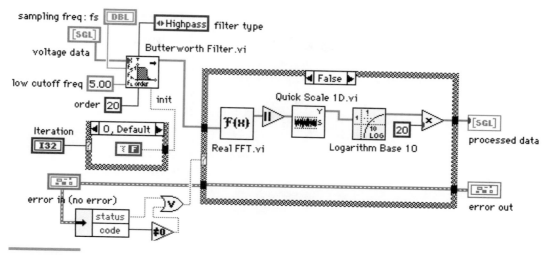

Figure D3.10(d)

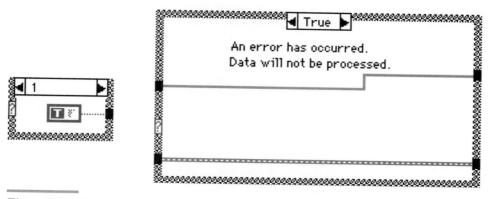

Figure D3.10(e)

Drill Problems
Day 4

Drill Problem 4.1

1. VIs to be used: **s081_AcqNScansDTrig.vi** (The VI is already complete.)
2. Objective: To learn how to perform digital triggered analog input data acquisition.
3. Estimated time: 10–15 minutes
4. Related chapter: Chapter 8
5. Key objects, VIs, and functions in this drill problem:
   VIs in **Functions >> Data Acquisition >> Analog Input**
6. Instructions
   Open the VI and examine its sub VIs, and different parameters for digital triggered analog input data acquisition. Read the information under **Windows >> Show VI Info** . . . . (You can also use the shortcut **Ctrl-I** to bring up the information.) The input channel(s) must be listed in reverse order and channel 0 must be included for multichannel scanning using LabPC-1200.

   For more information about the VI, refer to the example **Acquire N Scans Digital Trig.vi**, which can be found in LabVIEW directory **>> examples >> daq >> anlogin >> anlogin.llb**.

Drill Problem 4.2

1. VIs to be used: **s082_AcqNScansExtChClk.vi** (The VI is already complete.)
2. Objective: To learn how to perform analog input data acquisition with an external channel clock.
3. Estimated time: 10–15 minutes
4. Related chapter: Chapter 8
5. Key objects, VIs, and functions in this drill problem:
   VIs in **Functions >> Data Acquisition >> Analog Input**
6. Instructions
   Open the VI and examine its sub VIs, and different parameters for analog input data acquisition with an external channel clock. Read the information under **Windows >> Show VI Info** . . . . (You can also use the

shortcut **Ctrl-I** to bring up the information.) The input channel(s) must be listed in reverse order and channel 0 must be included for multi-channel scanning using LabPC-1200.

For more information about the VI, refer to the example **Acquire N Scans - ExtChanClk.vi**, which can be found in LabVIEW directory >> **examples >> daq >> anlogin >> anlogin.llb**.

Drill Problem 4.3

1. VIs to be used: **s083_ContAcq&GphExtScanClk.vi** (The VI is already complete.)
2. Objective: To learn how to perform continuous analog input data acquisition with an external scan clock.
3. Estimated time: 10–15 minutes
4. Related chapter: Chapter 8
5. Key objects, VIs, and functions in this drill problem:
   VIs in **Functions >> Data Acquisition >> Analog Input**
6. Instructions

   This example uses a **Waveform Graph** for continuous data display, but the textbook states that **Waveform Graph**s are for offline display. Such a statement is true, but **Waveform Graph**s can still be used for online display, and **Waveform Chart**s for offline display, too. In this drill problem, you will not be able to use the memory of **Waveform Chart**s during the online display. Consequently, **Waveform Graph**s will refresh the display at each iteration because they do not have memory.

   Open the VI and examine its sub VIs, and different parameters for continuous analog input data acquisition with an external scan clock. Read the information under **Windows >> Show VI Info ....** (You can also use the shortcut **Ctrl-I** to bring up the information.) The input channel(s) must be listed in reverse order and channel 0 must be included for multichannel scanning using LabPC-1200.

   For more information about the VI, refer to the example **Cont Acq&Graph ExtScanClk.vi**, which can be found in LabVIEW directory >> **examples >> daq >> anlogin >> anlogin.llb**.

Drill Problem 4.4

1. VIs to be used: **p101_DigWrt&AI.vi** and **p101_AISmplChs.vi** (Templates provided.)
2. Objective: To learn how to use high-level data acquisition VIs.
3. Estimated time: 30–35 minutes
4. Related chapter: Chapter 10
5. Key objects, VIs, and functions in this drill problem:
   Functions >> Data Acquisition >> Analog Input >> AI Sample Channels.vi
   Functions >> Data Acquisition >> Digital I/O >> Write to Digital Port.vi
6. Instructions

   This drill problem uses a sub VI **p101_AISmplChs.vi**, which is a modified version of the high-level analog input VI, **AI Sample Channels.vi**. Open the template VI **p101_AISmplChs.vi**, create a new terminal for the input **iteration (init:0)**, and save the change. Now, open the template VI **p101_DigWrt&AI.vi**, and complete it as shown in Figure D4.4. Read the information using the shortcut **Ctrl-I** for signal connection. Note that this drill problem uses another high-level digital I/O VI, **Write to Digital Port.vi**.

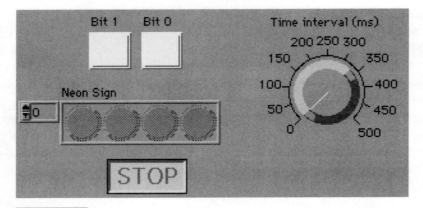

Figure D4.4(a)

Drill Problems • Day 4

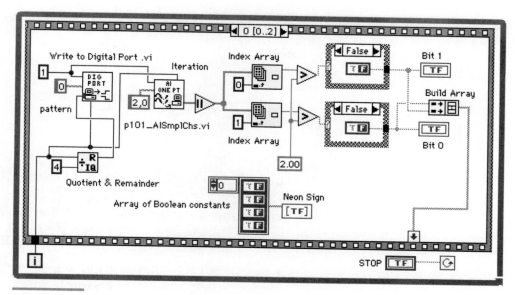

Figure D4.4(b)

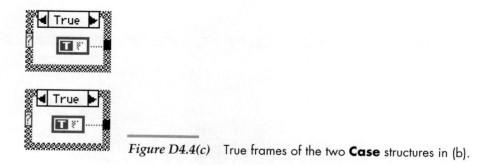

Figure D4.4(c) True frames of the two **Case** structures in (b).

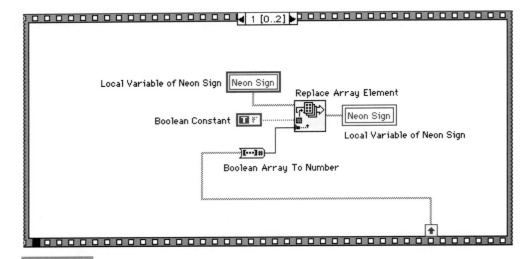

Figure D4.4(d)

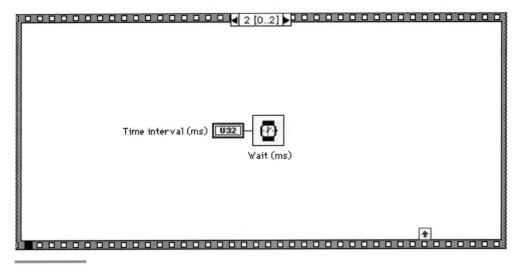

Figure D4.4(e)

Drill Problem 4.5

1. VIs to be used: **p111_WrtRdBin1DCts.vi** (Template provided.)
2. Objective: To learn how to write a binary data file continuously.
3. Estimated time: 20–25 minutes
4. Related chapter: Chapter 11
5. Key objects, VIs, and functions in this drill problem:
   VIs in **Functions >> File I/O**
6. Instructions

   Complete the VI as shown in Figure D4.5. Frame 0 saves the integer coming from the *iteration terminal* of the **While Loop** in binary format every 1 second until either an error occurs or the button **QUIT** is pressed. Then frame 1 reads the binary data back from the file and displays them in **data** in the front panel. View the file using a text editor program to see if the data are understandable.

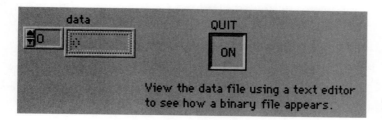

Figure D4.5(a)

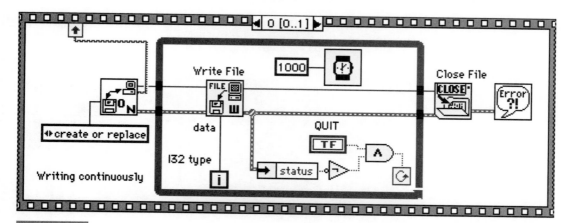

Figure D4.5(b)

Drill Problems • Day 4

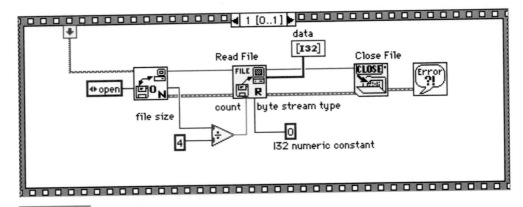

Figure D4.5(c)

Drill Problem 4.6

1. VIs to be used: **p112_WrtRdBin2D.vi** (Template provided.)
2. Objective: To learn how to write and read a 2-D binary data file without using header information.
3. Estimated time: 20–25 minutes
4. Related chapter: Chapter 11
5. Key objects, VIs, and functions in this drill problem:
   VIs in **Functions >> File I/O**
6. Instructions
   Complete the VI as shown in Figure D4.6. This drill problem shows how to write and read 2-D binary data without using header informa-

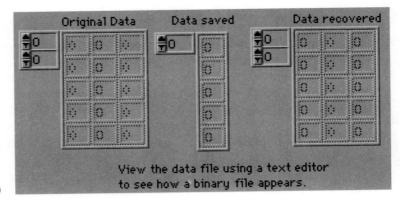

Figure D4.6(a)

tion. The number of columns in the 2-D binary data is determined by the number of channels, which is assumed to be known.

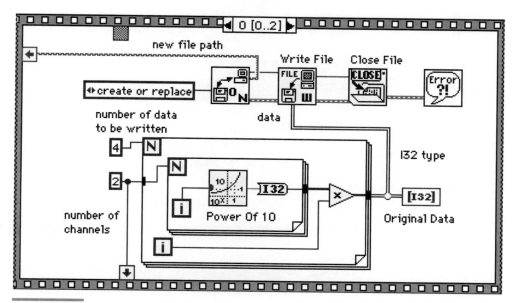

Figure D4.6(b)

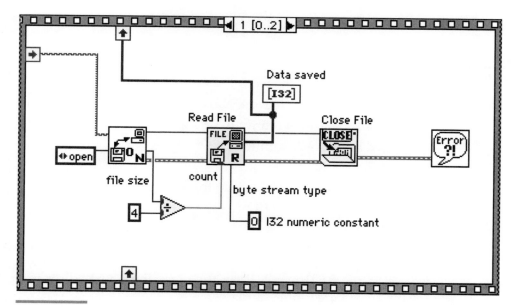

Figure D4.6(c)

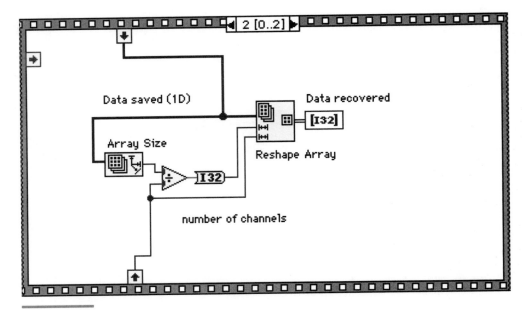

Figure D4.6(d)

Drill Problem 4.7

1. VIs to be used: **p113_WrtRdASCIICts.vi** (Template provided.)
2. Objective: To learn how to write ASCII data to a file continuously.
3. Estimated time: 20–25 minutes
4. Related chapter: Chapter 11
5. Key objects, VIs, and functions in this drill problem:
   VIs in **Functions >> File I/O**
6. Instructions
   Complete the VI as shown in Figure D4.7. This drill problem writes ASCII data to a file continuously, reads them back, and displays them in **data** in the front panel. View the file using a text editor program to see if the data are understandable.

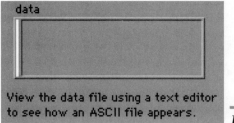

Figure D4.7(a)

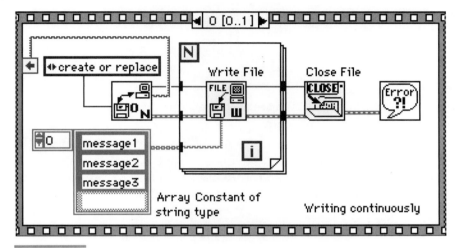

Figure D4.7(b)

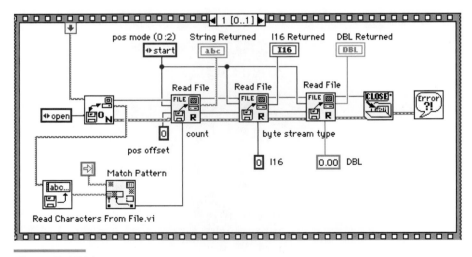

Figure D4.7(c)

Drill Problem 4.8

1. VIs to be used: **p114_WrtRdMixed.vi** (Template provided.)
2. Objective: To learn how to save data in both binary and ASCII format to a file.
3. Estimated time: 20–25 minutes
4. Related chapter: Chapter 11
5. Key objects, VIs, and functions in this drill problem:
   VIs in **Functions >> File I/O**
6. Instructions

   Complete the VI as shown in Figure D4.8. This drill problem shows how to save ASCII and binary data to a single file together, so that you can view and understand the ASCII data while reducing the file size by the rest of the binary data. In the front panel, the display mode '\' **Codes Display** is chosen for **String Input** and **String Returned** in order to show the accuracy in the returned text data. After saving the mixture of data to a file, view the data using a text editor to see which part of the data is understandable.

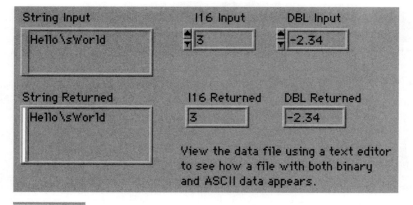

Figure D4.8(a)

Drill Problems • Day 4

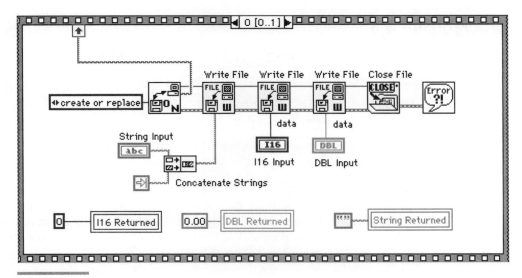

Figure D4.8(b)

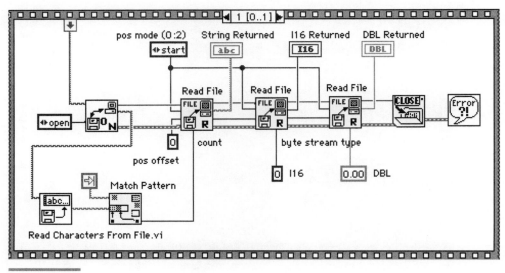

Figure D4.8(c)

Drill Problem 4.9

1. VIs to be used: **p115_WrtRdDatalog.vi** (Template provided.)
2. Objective: To learn how to save and read data in clusters.
3. Estimated time: 20–25 minutes
4. Related chapter: Chapter 11
5. Key objects, VIs, and functions in this drill problem:
   VIs in **Functions >> File I/O**
6. Instructions
   Complete the VI as shown in Figure D4.9. This drill problem shows how to save and read data in clusters. Note that you can retrieve the data correctly by this method only in LabVIEW since the data structure *cluster* is valid only in LabVIEW. Therefore, you will not be able to open the file created by this drill problem without using LabVIEW. (This is not an accurate statement since the cluster data can still be recovered by other programming languages if how LabVIEW saves clusters is known. However, such a case is not assumed both in this manual and in the main textbook.)

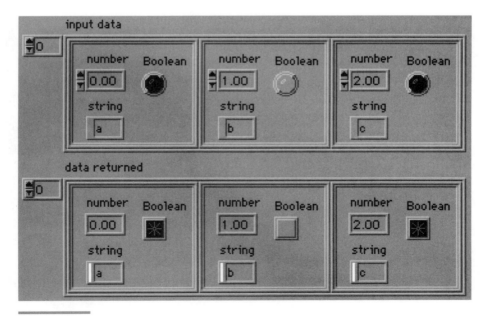

Figure D4.9(a)

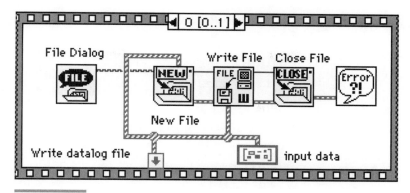

Figure D4.9(b)

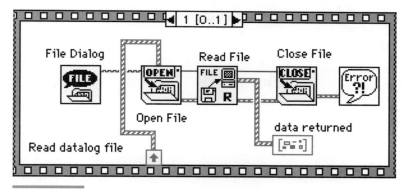

Figure D4.9(c)

Drill Problem 4.10

1. VIs to be used: **p116_RtrvDtalogDtaHalo.vi** (Template provided.)
2. Objective: To learn how to save and read data without using file I/O VIs.
3. Estimated time: 20–25 minutes
4. Related chapter: Chapter 11
5. Key objects, VIs, and functions in this drill problem:
   Functions >> Time & Dialog >> Seconds to Data/Time
6. Instructions

    Step 1: Open **s116_GenDtalogDta.vi**, and choose **Log at Completion** from the pull-down menu **Operate**. Run the VI multiple times to generates a datalog file. Such a datalog file also stores time information, which can be retrieved by Step 5.

Step 2: Open **p116_RtrvDtalogDtaHalo.vi**, and go to the diagram window. The sub VI **s116_GenDtalogDta.vi** is already placed in it.

Step 3: Right click on the sub VI and select **Enable Database Access** from its pop-up menu.

Step 4: Finish the wiring as shown in Figure D4.10.

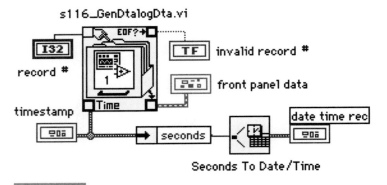

Figure D4.10

Step 5: Run the VI multiple times while changing the value in **record #**. A series of snapshots of data will be retrieved.

Step 6: If the sub VI has lost the connection with its data file that you created in Step 1, you can reconnect them by selecting **Operate >> Data Logging >> Change Log File Binding . . .** , and link to the correct datalog file again.

Drill Problem 4.11

1. VIs to be used: **p117_SvDataWithPref.vi** (Template provided.)
2. Objective: To learn how to save parameters upon quitting a VI and recover them from the VI's preferences file when reopening the VI.
3. Estimated time: 25 minutes
4. Related chapter: Chapter 11

5. Key objects, VIs, and functions in this drill problem:
   VIs in **Functions >> File I/O**
   Functions >> String >> Array To Spreadsheet String
6. Instructions

   This drill problem assumes that there exists a text file **preferences.txt** in the *same* directory where the VI is saved. Complete the VI as shown in Figure D4.11.

   The functionality of the VI is the following: When you quit the VI by pressing **QUIT**, the values in **x**, **y**, and **z** will be saved in the file **preferences.txt**, which is assumed to exist in the same directory where the VI is saved. When you run the VI next time, it will load the values of **x**, **y**, and **z** from **preferences.txt** and display them in each control. In order to

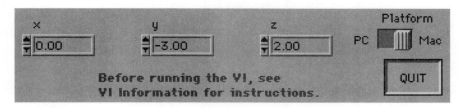

Figure D4.11(a)

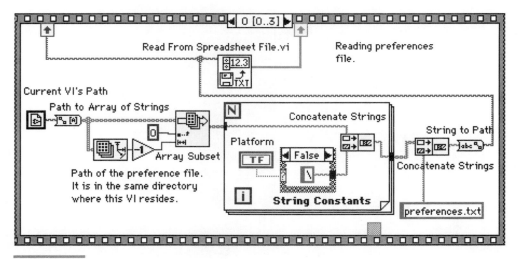

Figure D4.11(b)

see the change in the values, change them to a number that is different from the saved values before running the VI again.

Figure D4.11(c)   True frame of the **Case** structure in (b).

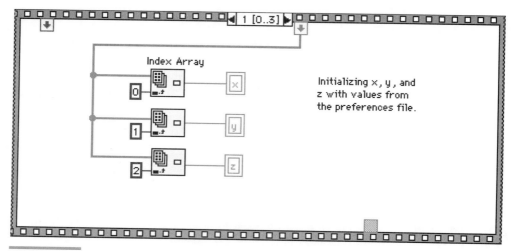

Figure D4.11(d)

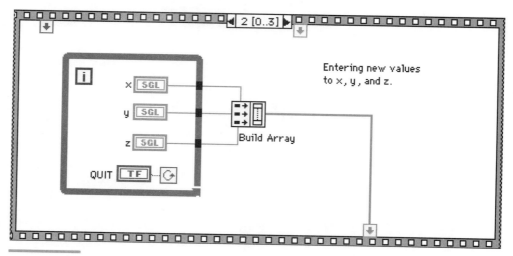

Figure D4.11(e)

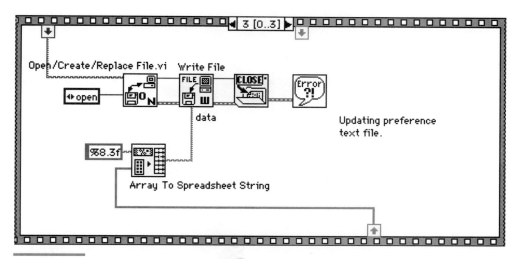

Figure D4.11(f)

Drill Problem 4.12

1. VIs to be used: **p121_ScrllbarCtrl.vi** (Template provided.)
2. Objective: To learn how to scroll the text in the string indicator as its window becomes full.
3. Estimated time: 15–20 minutes
4. Related chapter: Chapter 12
5. Key objects, VIs, and functions in this drill problem:
   Controls >> String & Table >> String Indicator
   Functions >> String >> Format Into String
6. Instructions

   This drill problem shows how you can scroll the text window of **String Indicator** when it becomes full of text using Attribute Nodes. (See Figure D4.12.) In the False frame of the **Case** structure, you simply connect

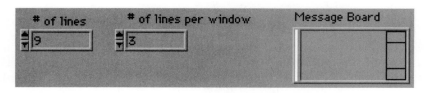

Figure D4.12(a)

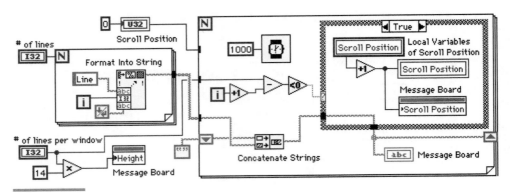

Figure D4.12(b)

the output of the **Concatenate Strings** to the string indicator **Message Board** through the False frame.

Drill Problem 4.13

1. VIs to be used: **s131_LV&Serl.vi** (The VI is already complete.)
2. Objective: To learn how to perform the self-loop test for serial communication.
3. Estimated time: 10 minutes
4. Related chapter: Chapter 13
5. Key objects, VIs, and functions in this drill problem:
   VIs in **Functions >> Instrument I/O >> Serial**
6. Instructions
   Open the VI and read the VI information using **Ctrl-I**. Follow the instructions there.
   For more information about the VI, refer to the example **LabVIEW <-> Serial.vi**, which can be found in LabVIEW directory **>> examples >> instr >> smplserl.llb**.

Drill Problem 4.14

1. VIs to be used: **s132_LV&GPIB.vi** (The VI is already complete.)
2. Objective: To learn how to test the connection to a GPIB instrument.
3. Estimated time: 10 minutes
4. Related chapter: Chapter 13
5. Key objects, VIs, and functions in this drill problem:
   VIs in **Functions >> Instrument I/O >> GPIB**
6. Instructions
   This drill problem is useful for testing the GPIB connection. First, run the VI to write to your GPIB device, and run it to read from the device. If the correct response comes back, the connection has been positively verified. Otherwise, you should check your device, cable, and the GPIB board.
   For more information about the VI, refer to the example **LabVIEW <-> GPIB.vi**, which can be found in LabVIEW directory **>> examples >> instr >> smplgpib.llb**.

Drill Problem 4.15

1. VIs to be used: **p141_AcqNScns1Ch_QS1D&FFT.vi** (Template provided.)
2. Objective: To learn how to perform a frequency analysis using FFT VIs.
3. Estimated time: 15 minutes
4. Related chapter: Chapter 14
5. Key objects, VIs, and functions in this drill problem:
   Functions >> Analysis >> Digital Signal Processing >> Real FFT.vi
   Functions >> Analysis >> Array Operations >> Quick Scale 1D.vi
6. Instructions
   Open the template, and complete it as shown in Figure D4.15. This drill problem uses **s141_QckScle1D&FFT.vi**, which returns the result of FFT using **Quick Scale 1D.vi** to normalize the output of **Real FFT.vi**. The sub VI **s141_QckScle1D&FFT.vi** is already complete and provided for

your convenience. Open it and study the diagram to see how FFT VIs are used.

A **Waveform Graph** is used to display the frequency content with **Bundle** to specify the x-interval, which is the sampling frequency divided by the total number of samples. Note that this drill problem is for a single-channel input.

For more information about the VI, refer to the example **Acquire N Scans.vi**, which can be found in LabVIEW directory >> **examples** >> **daq** >> **anlogin** >> **anlogin.llb**.

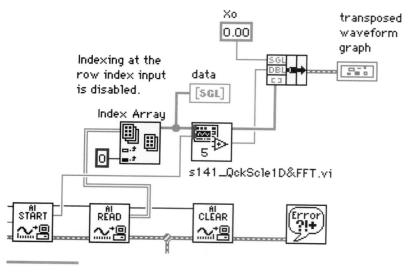

Figure D4.15

Drill Problems
Day 5

Drill Problem 5.1

1. VIs to be used: **s141_FreqResp.vi** (The VI is already complete.)
2. Objective: To learn how to select different types of digital filters, and how to use VIs for FFT.
3. Estimated time: 10–15 minutes
4. Related chapter: Chapter 14
5. Key objects, VIs, and functions in this drill problem:
   Functions >> Analysis >> Signal Generation >> Impulse Pattern.vi
   Functions >> Analysis >> Digital Signal Processing >> Real FFT.vi
   Functions >> Analysis >> Filters >> Butterworth Filter.vi, **Chebyshev Filter.vi**, and **Inverse Chebyshev Filter.vi**
   Functions >> Analysis >> Array Operations >> Quick Scale 1D.vi
6. Instructions
   Open the VI and examine how **Real FFT.vi** and **Quick Scale 1D.vi** are used. You can use this example to pick the optimal filter for your application. To see the frequency response of other filters, you can replace the three filters with other types.

Drill Problem 5.2

1. VIs to be used: **p142_FFTPairTest.vi** (Template provided.)
2. Objective: To learn how to use **Real FFT.vi** and **Inverse Real FFT.vi**.
3. Estimated time: 10–15 minutes
4. Related chapter: Chapter 14
5. Key objects, VIs, and functions in this drill problem:
   Functions >> Analysis >> Signal Generation >> Sinc Pattern.vi
   Functions >> Analysis >> Digital Signal Processing >> Real FFT.vi and **Inverse Real FFT.vi**
   Functions >> Analysis >> Array Operations >> Quick Scale 1D.vi

6. Instructions

    Complete the VI as shown in Figure D5.2. This drill problem generates a sinc function based on the two input parameters **delay** and **Δt**, displays it in the frequency domain via **Real FFT.vi**, and shows the recovered sinc function via **Inverse Real FFT.vi**. The ideal shape of a sinc function in the frequency domain is a square.

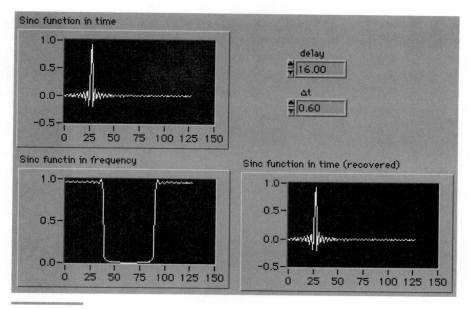

Figure D5.2(a)

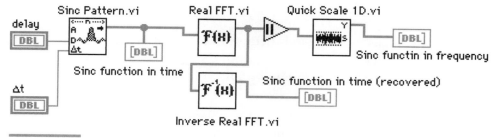

Figure D5.2(b)

Drill Problem 5.3

1. VIs to be used: **s143_DecIntTest.vi** (The VI is already complete.)
2. Objective: To see the effect of decimation and interpolation.
3. Estimated time: 10–15 minutes
4. Related chapter: Chapter 14
5. Key objects, VIs, and functions in this drill problem:
   LabVIEW directory **>> examples >> daq >> anlogin >> anlogin.llb >> Cont Acq&Chart (buffered).vi**
6. Instructions

   The effect of interpolation is D-fold periodic repetition of the original signal spectrum, and that of decimation is the stretch of the original signal spectrum by a factor of D, where D is either the interpolation factor or the decimation factor. Try different frequencies of input signal to witness the effect.

Drill Problem 5.4

1. VIs to be used: **s144_AliasTest&LPF.vi** (The VI is already complete.)
2. Objective: To see the effect of lowpass filtering in the time and the frequency domains.
3. Estimated time: 10–15 minutes
4. Related chapter: Chapter 14
5. Key objects, VIs, and functions in this drill problem:
   LabVIEW directory **>> examples >> daq >> anlogin >> anlogin.llb >> Cont Acq&Chart (buffered).vi**
6. Instructions

   This drill problem shows the effect of lowpass filtering. Try a square pulse train from a Function Generator at low frequency such as 1 or 2 Hz to see the effect clearly in time, and at high frequency to see the effect in frequency.

Drill Problem 5.5

1. VIs to be used: **p145_FittingPrac.vi** (Template provided.)
2. Objective: To learn how to use data fitting VIs.
3. Estimated time: 20–25 minutes
4. Related chapter: Chapter 14
5. Key objects, VIs, and functions in this drill problem:
   Functions >> Analysis >> Curve Fitting >> Linear Fit.vi and **Exponential Fit.vi**
   Functions >> Numeric >> Logarithmic >> Exponential
6. Instructions

   Open the template, and complete it as shown in Figure D5.5. This drill problem shows how to interpolate given data samples. This is useful when you need to estimate data sample values between the given samples.

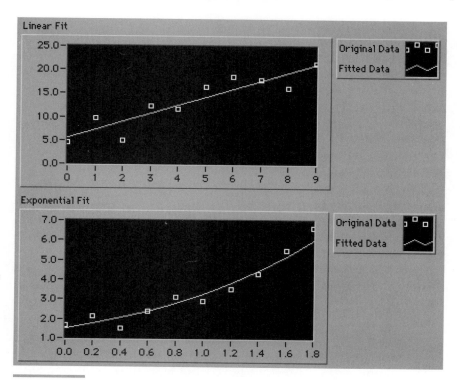

Figure D5.5(a)

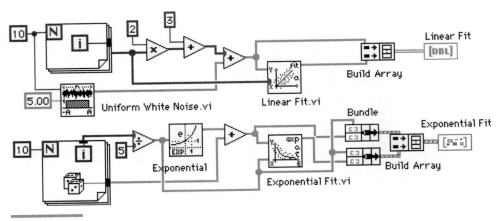

Figure D5.5(b)

Drill Problem 5.6

1. VIs to be used: **p151_While&Occrnce.vi** and **p151_WhileWoOccrnce.vi** (Templates provided.)
2. Objective: To learn how to use occurrences.
3. Estimated time: 25–30 minutes
4. Related chapter: Chapter 14
5. Key objects, VIs, and functions in this drill problem:
   Functions >> Advanced >> Synchronization >> Occurrences >> Generate Occurrence, **Wait on Occurrence**, and **Set Occurrence**
6. Instructions

   This drill problem shows the effect of occurrences. If you use occurrences VIs, they will stop the VI immediately even though the time delays in multiple **While Loop**s are different. Without the occurrences VIs, it will take the longest time delay to stop multiple **While Loop**s.

   Open the template VIs, and complete them as shown in Figure D5.6. The front panels of both template VIs are identical. The diagram window without occurrences VIs is Figure D5.6(b), and that with occurrences VIs is Figure D5.6(c).

Drill Problems • Day 5

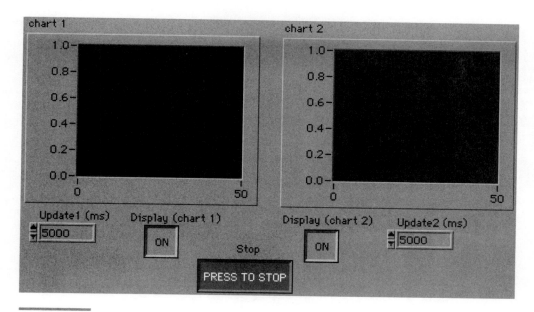

Figure D5.6(a)

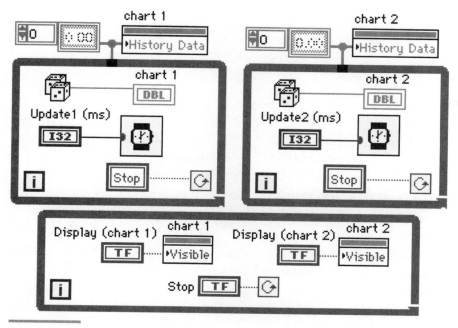

Figure D5.6(b)

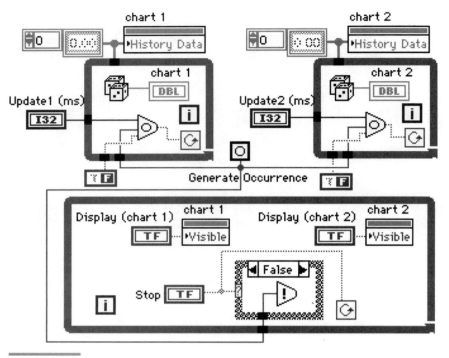

Figure D5.6(c)
The True frame of the **Case** structure is empty.

Drill Problem 5.7

1. VIs to be used:
   p152_ErrHndler_Main.vi (Template provided.)
   p152_ErrHndler_Proc1.vi (Template provided.)
   p152_ErrHndler_Proc2.vi (Template provided.)
   p152_ErrHndler_Proc3.vi (Template provided.)
   s152_ErrHndler_ErrDiply.vi (The VI is already complete.)
2. Objective: To review error tracking methods and learn how to design customized error handler VIs.
3. Estimated time: 25–30 minutes
4. Related chapter: Chapter 14
5. Key objects, VIs, and functions in this drill problem:
   N/A

Drill Problems • Day 5

6. Instructions

   This drill problem is to show how to design your own error handler VIs. Complete each VI as shown in Figure D5.7.

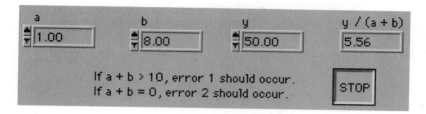

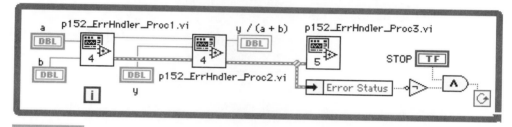

Figure D5.7(a)
Front panel and diagram window of **p152_ErrHndler_Main.vi**.

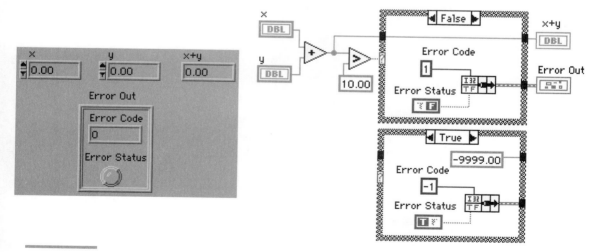

Figure D5.7(b)
Front panel and diagram window of **p152_ErrHndler_Proc1.vi**.

Drill Problems • Day 5

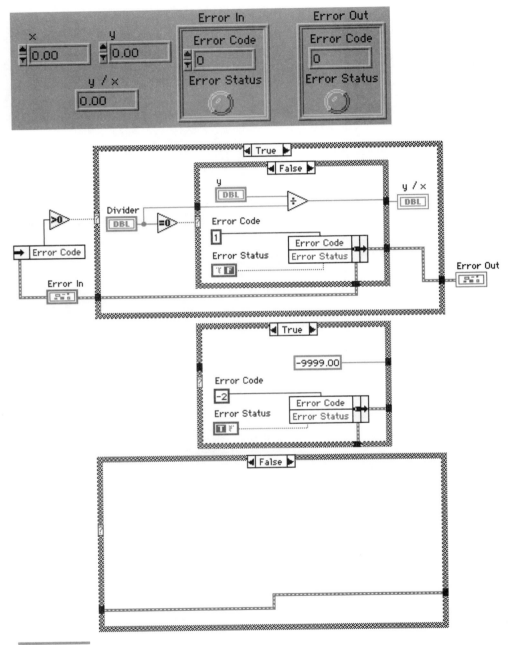

Figure D5.7(c)
Front panel and diagram window of **p152_ErrHndler_Proc2.vi**.

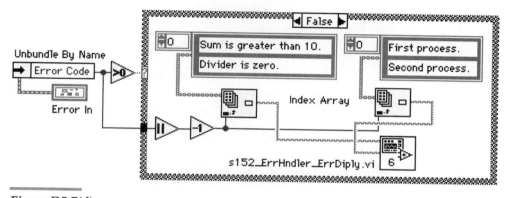

Figure D5.7(d)
Front panel and diagram window of **p152_ErrHndler_Proc3.vi**. The True frame of the **Case** structure is empty.

Drill Problems
Day 6

Drill Problem 6.1

1. VIs to be used: **p151_ListVIs.vi** (Template provided.)
2. Objective: To learn how to list VIs in the current directory and display the one that is double clicked in the list.
3. Estimated time: 10–15 minutes
4. Related chapter: Chapter 15
5. Key objects, VIs, and functions in this drill problem:
   Controls >> **List & Ring** >> **Single Selection Listbox**
   Functions >> **File I/O** >> **File Constants** >> **Current VI's Path**
   Functions >> **File I/O** >> **Strip Path**
   Functions >> **File I/O** >> **Advanced File Functions** >> **List Directory**
   Functions >> **Array** >> **Index Array**
6. Instructions
   Open the template VI, and complete it as shown in Figure D6.1. Notice the usage of two Attribute Nodes of **VI List** and the function **Wait (ms)**. The time value 200 msec is chosen to give enough time for you to double click a VI in **VI List**. This drill problem will display in **Selected VI** the name of the VI that is double clicked in **VI List** momentarily, and will be used for the next drill problem. Note that the output **stripped path** of **Strip Path** is wired to the input **directory path** of **List Directory**, and the output **file names** of **List Directory** is wired to the Attribute Node **Item Names** of **VI List**.

   The mechanical action setting of **QUIT** is set to **Latch When Released**, and its default value has been changed to TRUE. Those changes will prepare the VI to be ready to run again after each stop.

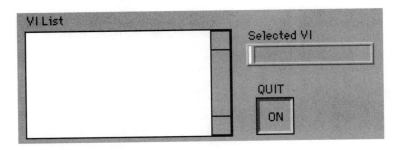

Figure D6.1(a)

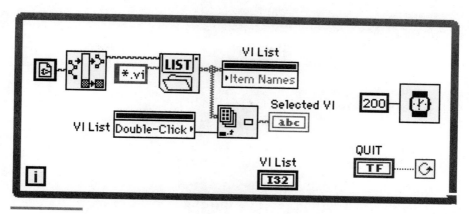

Figure D6.1(b)

Drill Problem 6.2

1. VIs to be used: **p152_DynamicLoad.vi** (Template provided.)
2. Objective: To learn how to load and unload sub VIs dynamically to use the memory more efficiently.
3. Estimated time: 20–25 minutes
4. Related chapter: Chapter 15
5. Key objects, VIs, and functions in this drill problem:
   Controls >> List & Ring >> Single Selection Listbox
   Functions >> File I/O >> File Constants >> Current VI's Path
   Functions >> File I/O >> Strip Path and **Build Path**
   Functions >> File I/O >> Advanced File Functions >> List Directory
   Functions >> Array >> Index Array
   Functions >> Application Control >> Open VI Reference
   Functions >> Application Control >> Call By Reference Node
   Functions >> Application Control >> Close Application or VI Reference
6. Instructions

   Open the template VI, and complete it as shown in Figure D6.2. This example is an extension of that in Drill Problem 6.1. Note that the output **stripped path** of **Strip Path** is wired to both the input **directory path** of **List Directory** and the input **base path** of **Build Path**, and its output **appended path** is wired to the input **vi path** of **Call By Reference Node**.

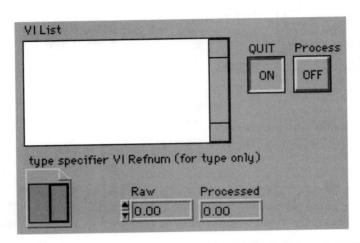

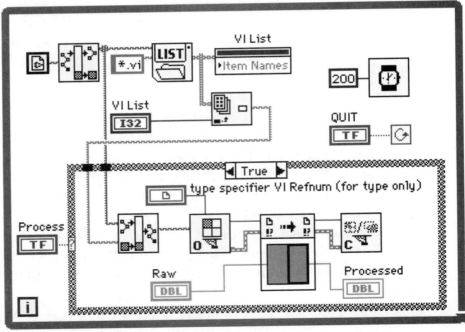

Figure D6.2
The False frame of the **Case** structure is empty.

This example will display the VIs in the currently directory in **VI List** and call dynamically the VI that is highlighted in the list when **Process** is pressed. The mechanical action of both **QUIT** and **Process** is **Latch When Released**, and their default values are TRUE and FALSE, respectively.

The control **type specifier VI Refnum (for type only)** is created by right clicking on the terminal and choosing **Create Control**. Then go to the front panel, right click on the control, and choose **Select VI Server Class >> Browse**. Navigate the directory to select **s151_AddOne.vi**. This will let LabVIEW know of the type of sub VIs to be called dynamically. Once the example is completed, run the VI while selecting different sub VIs. For the sake of convenience, two sub VIs, **s151_AddOne.vi** and **s151_SubtractOne.vi**, are provided in the solutions folder. Therefore, place a copy of each sub VI in the same directory where **p152_Dynamic Load.vi** is saved in order to have the two sub VIs available in the list. If you would like to add your own sub VIs to the list, make sure that they have the same structure as the two sub VIs provided: one input and one output. Otherwise, you should define the setting of **type specifier VI Refnum (for type only)** again.

Drill Problem 6.3

1. VIs to be used: **p153_init.vi**, **p153_main.vi**, **p153_menu1.vi**, and **p153_menu2.vi** (Templates provided.)
2. Objective: To learn how to create a professional startup screen of an application.
3. Estimated time: 40–45 minutes
4. Related chapter: Chapter 15
5. Key objects, VIs, and functions in this drill problem:
   Functions >> File I/O >> File Constants >> Current VI's Path
   Functions >> File I/O >> Strip Path and **Build Path**
   Functions >> Application Control >> Open VI Reference
   Functions >> Application Control >> Property Node
   Functions >> Application Control >> Close Application or VI Reference

6. Instructions

Open the template VIs, and complete them as shown in Figure D6.3. You will create a simple application with a professional startup screen. The functionality of the application is as follows: The initial message **"My First Application Version 1.0"** will appear on the screen for 3 seconds when you launch the application. Then the initial message disappears, and the main window takes it over. The main window has two links for menu 1 and menu 2. Menu 1 adds all of the elements in the data array and returns the result when you come back to the main window. Menu 2 multiplies all of the elements in the data array and returns the result when you come back to the main window. If you quit in the main window, the entire application terminates.

The following is the list of assumptions and the settings of each sub VI:

(a). All of the VIs are assumed to be stored in the same directory.

(b). The VI **p153_init.vi** has only the following settings chosen in its **VI Setup ...** window: **Run When Opened** in the **Execution Options** window, and **Auto-Center** in the **Window Options** window.

(c). All of the other VIs have only the following settings chosen in their **VI Setup ...** window: **Show Front Panel When Called** and **Close Afterwards if Originally Closed** in the **Execution Options** window, and **Auto-Center** in the **Window Options** window.

(d). The mechanical action of **Menu 1**, **Menu 2**, and **QUIT** in **p153_main.vi** is set to **Latch When Released**. The default values of them are FALSE, FALSE, and TRUE, respectively.

(e). The mechanical action of **Back to Main Window** in both **p153_menu1.vi** and **p153_menu2.vi** is **Latch When Released**, and the default value is TRUE for both.

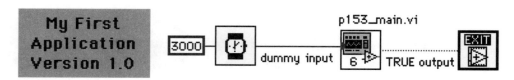

Figure D6.3(a)
Front panel and diagram window of **p153_init.vi**.

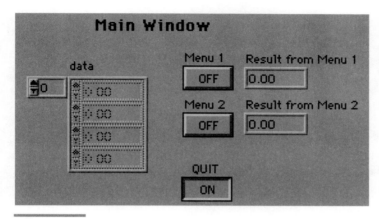

Figure D6.3(b)
Front panel of **p153_main.vi**.

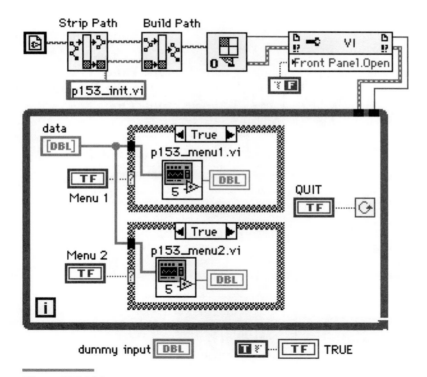

Figure D6.3(c)
Diagram window of **p153_main.vi**. The False frames of the two **Case** structures are empty.

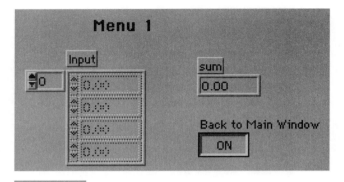

Figure D6.3(d)
Front panel of **p153_menu1.vi**.

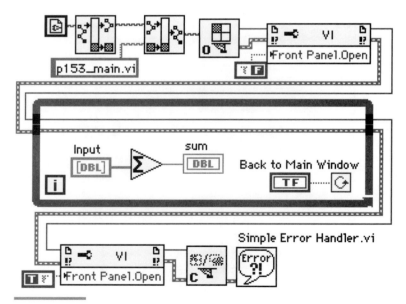

Figure D6.3(e)
Diagram window of **p153_menu1.vi**.

Drill Problems • Day 6

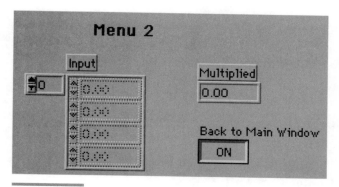

Figure D6.3(f)
Front panel of **p153_menu2.vi**.

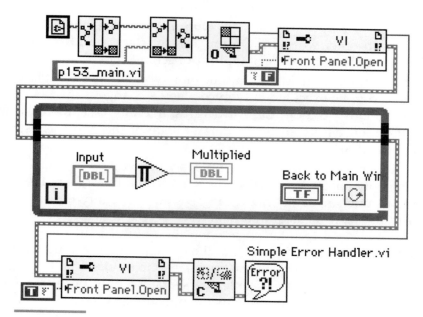

Figure D6.3(g)
Diagram window of **p153_menu2.vi**.

Drill Problem 6.4

1. VIs to be used: **p154_StatCheck.vi** (Template provided.)
2. Objective: To realize an algorithm that checks the status change continuously.
3. Estimated time: 10–15 minutes
4. Related chapter: Chapter 15
5. Key objects, VIs, and functions in this drill problem:
   Functions >> Boolean >> And
   Functions >> Comparison >> Not Equal?
   Functions >> Time & Dialog >> Wait (ms)
6. Instructions
   Open the template VI, and complete it as shown in Figure D6.4. This example flickers two LEDs **Entered** and **Left** when **Option 1** is changed. If the change is from FALSE to TRUE, the LED **Entered** will flicker once, executing its True frame only once. If the change is from TRUE to FALSE, the other LED **Left** will flicker once, executing its True frame only once. However, the state of **Option 1** remains that of the latest selection. In other words, if **Option 1** is selected, the LED **Entered** will flicker once, and the state of **Option 1** becomes TRUE.

   To use this example in your application, you will need to place the VIs that need to be executed only once when a change is made in the True frame of either or both **Case** structures. In either case, the False frames must be left empty since the default state of the two **Case** structures will be FALSE.

   The importance of such an example is when you need to take a certain action only once when a selection is made, but the state of the selection remains unchanged so that it can be used to identify the current state. For example, if the acquired data have been selected to be lowpass filtered, you need to filter once when the request is made, but an LED should remain lit indicating that the current data are a filtered version. Also, if you have a series of test sequences to be executed upon the user selection, you need to execute the sequence only once, not continuously, when such a request is made, but should keep a Boolean indicator on indicating that a test has been performed.

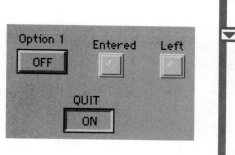 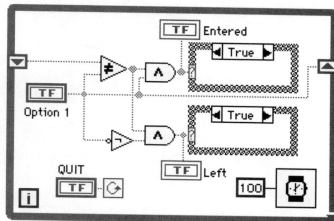

Figure D6.4
Both True and False frames of the two **Case** structures are empty. Note that the False frames must be empty since the default state of the two **Case** structures will be FALSE, unless you want to run some VIs repeatedly.

Drill Problem 6.5

1. VIs to be used: **p155_4ToggleSW.vi** (Template provided.)
2. Objective: To learn the general algorithm of alternative switching mechanism.
3. Estimated time: 15–20 minutes
4. Related chapter: Chapter 15
5. Key objects, VIs, and functions in this drill problem:
   Functions >> Comparison >> Not Equal?
6. Instructions

   This example shows a general case of realizing the alternative switching algorithm with four option choices. Open the template VI, and complete it as shown in Figure D6.5. Note that simply writing the opposite Boolean value of the latest selection to the rest of the options will not work but will create an infinite flip-flop case.

   The diagram window in Figure D6.5(b) shows how to achieve such a task with four option choices. A conditional write of the opposite

Boolean value to three **Local Variable**s is shown in each of the four innermost **Case** structures. Therefore, more **Local Variable**s will be necessary as the number of option choices increases, while the basic structure remains identical. Such an example is useful when designing a customized panel for an alternative user input. Once an option is selected, it will maintain the TRUE state until another option is selected.

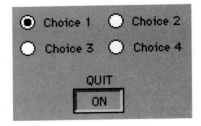

Figure D6.5(a)

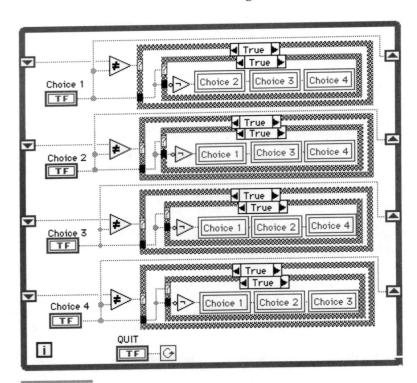

Figure D6.5(b)
The False frames of all **Case** structures are empty.

Drill Problem 6.6

1. VIs to be used: **p156_LocalVar&SR.vi**, **p156_AcquireData.vi**, and **p156_ProcData.vi** (Templates provided.)
2. Objective: To learn memory efficient programming using shift registers instead of **Local Variables**.
3. Estimated time: 20–25 minutes
4. Related chapter: Chapter 15
5. Key objects, VIs, and functions in this drill problem:
   Functions >> Structures >> While Loop and **Case** structure
   Functions >> Analysis >> Probability and Statistics >> Mean.vi
6. Instructions

   This example shows how to replace **Local Variables** by a **While Loop** and a **Case** structure. Open each template VI, and complete them as shown in Figure D6.6. The two Boolean buttons **Acquire** and **Process** have **Latch When Released** as their mechanical action settings.

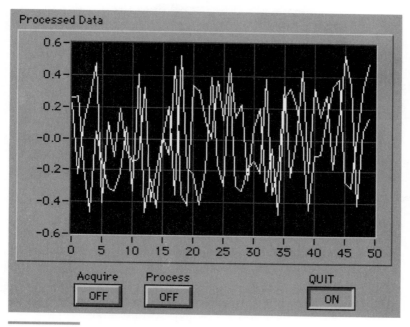

Figure D6.6(a)

The functionality of the example is as follows: Even though the two sub VIs **p156_AcquireData.vi** and **p156_ProcData.vi** are in the outer **While Loop**, they acquire or process the data only when either **Acquire** or **Process** is pressed. However, the two sub VIs return an empty data if **Acquire** or **Process** is not TRUE. In order to keep the valid data after acquiring or processing only once, a pair of shift registers and a **While Loop** are used as shown in Figure D6.6(b). Otherwise, two indicators to the outputs of the two **Case** structures need be created, and their **Local Variable**s should be used in the False frames of the two **Case** structures to complete the tunnels. Since the usage of **Local Variable**s should be minimized in order to maximize the efficiency in memory usage, this example shows how to keep the valid data without using **Local Variable**s.

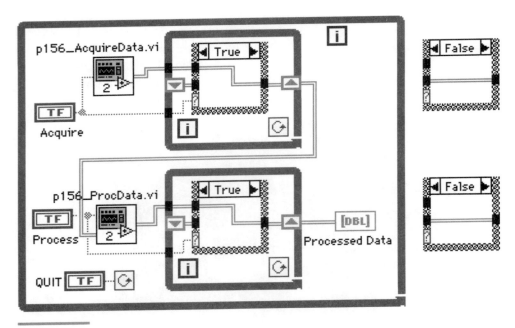

Figure D6.6(b)

Figure D6.6(c) Front panel of **p156_AcquireData.vi**.

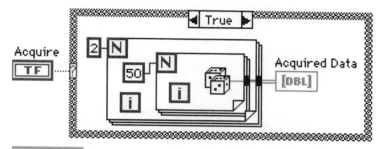

Figure D6.6(d)
Diagram window of **p156_AcquireData.vi**. The False frame is empty.

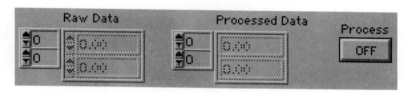

Figure D6.6(e)
Front panel of **p156_ProcData.vi**.

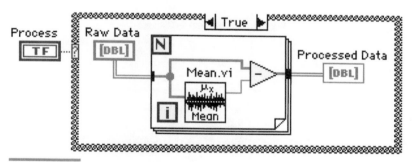

Figure D6.6(f)
Diagram window of **p156_ProcData.vi**. The False frame is empty.

Drill Problem 6.7

1. VIs to be used: **s156_LocalVar&SR.vi** (The VI is already complete.)
2. Objective: To learn how to save the top level VI and all of the sub VIs manually.

3. Estimated time: 5–10 minutes
4. Related chapter: Chapter 15
5. Key objects, VIs, and functions in this drill problem:

    N/A
6. Instructions

    Suppose that you want to transport all of the VIs associated with a top-level VI—in this example, **s156_LocalVar&SR.vi**. To collect all of the VIs, including the top-level one, **Save With Options . . .** under the pull-down menu **File** can be used, and this example explains how to achieve such a task.

    Step 1: Open the solution VI **s156_LocalVar&SR.vi**.

    Step 2: Select **Save With Options . . .** from the pull-down menu **File**. This will bring up the window shown in Figure D6.7.

    Step 3: Choose **Application Distribution**, then **No password change**. This will change the selection from **Application Distribution** to **Custom Save**.

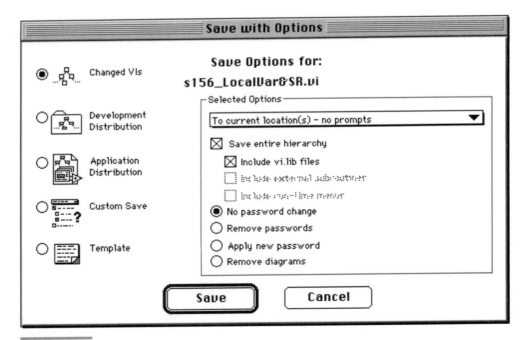

Figure D6.7

Step 4: Click **Save** to start collecting the VIs. This will bring up another window where you can specify the name of the destination file. Note that you can save all of the VIs either in a LabVIEW directory file **.llb** or in a directory as an individual VIs. Either way, LabVIEW will save all of the sub VIs, including the top-level VI itself, into the destination file or directory.

This feature is useful, especially if the number of sub VIs is large and the location of each sub VI is widely spread. To transport your application to a different machine, you can collect and save all of the sub VIs associated with the top-level VI and move the resulting **.llb** file or directory.

In this example, **s156_LocalVar&SR.vi** uses three sub VIs: **s156_AcquireData.vi**, **s156_ProcData.vi**, and **Mean.vi**. After completing the steps, LabVIEW will create either a **.llb** file or a directory where all of the three sub VIs are stored. Note that you must *not* select the option **Remove diagrams**; otherwise, LabVIEW will remove the diagrams of the sub VIs before saving them to make the resulting file size as small as possible. **Remove diagrams** is used when creating an executable.

Index

A

AcqNScns1Ch_QS1D&FFT, 67–68
Acquire
    Boolean buttons, 93–95
AcquireData, 93–95, 97
Acquire N Scans, 68
    LabVIEW directory example, 32
Acquire N Scans Digital Trig, 48
Acquire N Scans - ExtChanClk, 49
AI Sample Channels, 50
Algorithm
    alternative selection switching mechanism, 91–92
    change
        status continuous checks, 90–91
Aliasing
    frequency changing, 44
AliasTest, 42–45
AliasTest&LPF, 72
Alternative selection switching mechanism
    algorithm, 91–92
Analog input acquisition
    parameters, 32
    snapshot, 32

Analog input data acquisition
    buffered continuous, 32–33
    continuous, 49
        external scan clock, 49
    digital triggered, 48–49
    external channel clock, 48–49
Analog output
    one point, 33
Application
    startup screen creation, 85–89
Application Distribution, 96
Array
    build
        input settings, 23–24
    cluster, 24–25
    creating, 9–10
    data
        individual sample, 9–10
    data display two-D
        intensity graph, 26–29
    element
        replacement, 30
    extract, 24–25
    indexing, 11–12
    indexing elements, 29–30
    input, 24

Index

ASCII data
    saving and reading, 37
    string constant Testing Message, 37
    writing continuously, 56–57
ASCII format
    and binary format
        saving data, 58–59
Attribute node
    chart, 11–12
    string indicator, 65–66
    VI List, 82
    waveform chart, 8, 15
Attribute Node Item Names, 82
Auto indexing feature, 10

B

Binary data file
    continuously
        write, 53–54
    saving and reading, 35–36
    writing, 39
Binary format
    and ASCII format
        saving data, 58–59
    while loop, 50
Binary string data
    converting
        numeric data, 38–42
    reading, 39
Boolean buttons
    Acquire and Process, 93–95
Boolean switch
    creating, 5–7
    mechanical action, 11–12
    mechanical action settings, 18
    while loop, 22
    wiring technique, 20
Boolean value
    Local Variables, 92
Boundary
    for loop, 10
    while loop, 11–12
Buffered continuous analog input data
        acquisition, 32–33
Build array
    input settings, 23–24
Build Path, 83–85
Bundle
    frequency content, 68
    waveform chart, 11–12

C

Call By Reference Node, 84
Case structure, 8, 15, 39, 51f, 64f, 65–66
    error handling, 79f
    False frames, 87f, 92f
    StatCheck, 90–91
    True and False frames, 91f
    True frame, 76f
Change
    algorithm
        status continuous checks, 90–91
Channels
    vs. columns, 55
Charts, 7–8
    attribute node, 11–12
Clusters
    array, 24–25
    data
        saving and reading, 60–61
    indexing elements, 29–30
    updating elements, 30
Codes Display, 58
Columns
    vs. channels, 55
Concatenate Strings, 66
Conditional terminal, 20
Cont Acq&Chart (buffered), 33
Cont Acq&Graph Ext ScanClk, 49
Continuous analog input data acquisition
    buffered, 32–33
    external scan clock, 49
Cont Pulse Train, 35
Controls
    creating, 4–5
Count Events, 35
Create Control, 85
CtsAcq&Chart(buff), 34, 35, 43
Current directory, 82
Custom Save, 96

Index

D

2-D. *See* Two-D
Data
   acquisition VIs
      high level, 50–52
   array
      individual sample, 9–10
   clusters
      saving and reading, 60–61
   samples
      stream, 9–10
   saving and reading
      without I/O VIs, 60–61
DC generation, 33
Decimation
   effect, 72
DecIntTest, 72
Default state
   changing, 11–12
Delay parameters
   sinc functions, 71
D-fold periodic repetition, 72
Diagrams
   removal, 97
Diagram window
   completion, 11–12
Digital filters
   selection, 70
Digital triggered analog input data acquisition
   external channel clock, 48–49
   performance, 48–49
DigWrt&AI, 50
Display two-D array data
   intensity graph, 28–29
DynamicLoad, 83–85

E

Element
   input, 24
   orders, 25
Enable Database Access, 62
ErrHndler, 76–79
Error handling
   customizing, 76–79
Error tracking methods, 76
Event counting, 35
Execution option, 7
External channel locks
   analog input data acquisition, 48–49
External scan clock
   continuous analog input data acquisition, 49
Extract arrays, 24–25

F

False frame
   Case structure, 87f, 92f
   Message Board, 66
FFTPairTest, 70–71
FFTproc, 42–45
FFT VIs, 70
   frequency analysis
      performance, 67–68
File preference.txt
   saving
      Quit, 63
FittingPrac, 73
Flickers, 90–91
For loops, 10
FreqResp, 70
Frequency analysis
   capability, 43
   FFT VIs
      performance, 67–68
Frequency changing
   aliasing, 44
Frequency content
   Bundle, 68
Function Generator
   square pulse train, 72
   wiring technique, 44
Functions, 2–15

G

GenDtalogDta, 61, 62
Generate 1 Point on 1 Channel, 33
GPIB instrument
   connection test, 67

H

High level data acquisition VIs, 50–52

I

Indexing
    arrays, 11–12
    effect, 9–10
    elements
        cluster and arrays, 29–30
Indicators
    creating, 4–5
Init.vi, 85–89
Input
    acquisition analog
        buffered continuous, 32–33
        continuous, 49
        continuous external scan clock, 49
        digital triggered, 48–49
        external channel clock, 48–49
        parameters, 32
        snapshot, 32
    array, 24
    base path, 83–85
    channels
        addressing, 32
    data acquisition
        buffered continuous analog, 32–33
    directory path, 82, 83–85
    element, 24
    iteration
        terminal creation, 50–52
    parameters delay
        sinc functions, 71
    settings
        build array, 23–24
    string, 58
    vi path, 84
Intensity graph
    display two-D array data, 28–29
    two-D array data, 26–27
Interpolation
    data samples, 73
    effect, 72

Intiger
    saving, 53
Inverse Real FFT, 70–71
Iteration terminal
    creating new, 44
    saving, 53

L

LabPC-1200
    multichannel scanning, 32
    wiring technique, 44
LabVIEW
    directory
        examples, 32–35
    example modification, 43
    using built-in examples, 32–35
List Directory, 82, 83–85
ListVIs, 82
Local Variables
    Boolean value, 92
LocalVar&SR, 93–97
Log at Completion, 61
Lowpass filtering, 90–91
    effect, 72
LV&GPIB, 67
LV&Serl, 66

M

Main.vi, 85–89
Matrix
    building, 23–24
Mean, 97
Mechanical action settings
    Boolean switch, 18
Memory
    efficient programming, 93–95
    efficient use, 83–85
Menu1.vi, 85–89
Menu2.vi, 85–89
Message Board
    string indicator, 66
Moving Average (MA), 13–15

Index

Multichannel display
  waveform chart, 11–15
Multichannel scanning
  LabPC-1200, 32
Multiple vectors, 23–24

N

Numeric data, 38–42

O

Occurrences
  use and effect, 74
One-D binary data file
  saving and reading, 35–36
One point analog output, 33
Operate, 61
Output
  analog
    one point, 33
  appended path, 84
  file names, 82
  stripped path, 83–85

P

Parameters
  saving, 62–65
Pin connections, 35
Plotting
  multichannels of data, 18–20, 22
Preferences.txt
  text file, 63
ProcData, 93–95, 97
Process, 85
  Boolean buttons, 93–95
Professional startup screen
  creation, 85–89
Pulse train
  generating and monitoring, 35

Q

QckScle1D&FFT, 67–68
Quick Scale 1D, 67–68, 70
Quit, 85
  mechanical action settings, 82
  saves file preference.txt, 63

R

Random number display, 7–8
Real FFT, 67–68, 70–71
Real time, 7–8
Remove diagrams, 97
RtrvDtalogDtaHalo, 60–61, 62

S

Sampling frequency, 68
  creating new terminal, 44
Save
  custom, 96
  manual
    top level VIs and sub VIs, 95–97
Save With Options, 96
Screen
  startup creation, 85–89
ScrllbarCtrl, 65–66
Selected VI, 82
Selection
  state, 90–91
  switching mechanism
    algorithm, 91–92
Select VI Server Class, 85
Self loop test
  performance, 66
Sequence structure, 15, 39
Serial communication
  self loop test, 66
Setup options
  VIs, 5–7
Shift register, 11–12, 93–95
Shortcuts, 4
  for information, 32–35

Signal connection, 35
Sinc functions
    generation, 71
Square pulse train
    Function Generator, 72
Startup screen
    creation, 85–89
StatCheck, 90–91
String constant Testing Message
    ASCII data file, 37
String control, 5–7
String data
    reading, 39
String indicator
    Attribute Nodes, 65–66
    Message Board, 66
    scrolling text, 65–66
String Input, 58
String Returned, 58
Strip Path, 82–85
Stripped path, 82
Sub VIs
    assumption list, 86
    examining, 32–35
    manual save, 95–97
SvDataWithPref, 62–65

T

Testing Message
    string constant
        ASCII data file, 37
Text file
    preferences.txt, 63
Top level VIs
    manual save, 95–97
True frame
    Case structure, 76f
Tunnels, 10, 94
Two-D array data
    display
        intensity graph, 28–29
        intensity graph, 26–27
Two-D data files
    reading and writing, 54–56
Type specifier VI Refnum, 85

V

Vector
    building, 23–24
VI List
    Attribute Nodes, 82
VI Refnum, 85
VIs, 2–15
    building, 7–8
        complicated, 13–15
    completing, 10, 19, 24–25, 27, 30
    creating, 4–5
    dynamically
        load and unload, 83–85
    occurrences, 74
    opening, 5–7, 32–35
    preference file, 62–65
    running, 24
    setup options, 5–7

W

Wait(ms), 82
Waveform
    chart, 49
        attribute node, 8, 15
        multichannel display, 11–15
        plotting, 18–19
    generating and monitoring, 34
    graph, 49, 68
        plotting, 18–20
        vs. XY graph, 22
While loop, 5–7, 19, 20
    AliasTest, 43
    binary format, 50
    Boolean switch, 22
    boundary, 11–12
    Local Variables, 93–95
    occurrences, 74
    shift register, 94
While&Occrnce, 74
WhileWoOccrnce, 74
Window option, 7
Wiring technique, 2
    Boolean switch, 20

Index

Function Generator, 44
Write to Digital Port, 50
WrtRdASCIICts, 56–57
WrtRdBin2D, 54–56
WrtRdBin1DCts, 53–54
WrtRdDatalog, 60–61
WrtRdMixed, 58–59

X

XY graph
    vs. waveform graph, 22

LICENSE AGREEMENT AND LIMITED WARRANTY

READ THE FOLLOWING TERMS AND CONDITIONS CAREFULLY BEFORE OPENING THIS DISK PACKAGE. THIS LEGAL DOCUMENT IS AN AGREEMENT BETWEEN YOU AND PRENTICE-HALL, INC. (THE "COMPANY"). BY OPENING THIS SEALED DISK PACKAGE, YOU ARE AGREEING TO BE BOUND BY THESE TERMS AND CONDITIONS. IF YOU DO NOT AGREE WITH THESE TERMS AND CONDITIONS, DO NOT OPEN THE DISK PACKAGE. PROMPTLY RETURN THE UNOPENED DISK PACKAGE AND ALL ACCOMPANYING ITEMS TO THE PLACE YOU OBTAINED THEM FOR A FULL REFUND OF ANY SUMS YOU HAVE PAID.

1. **GRANT OF LICENSE:** In consideration of your payment of the license fee, which is part of the price you paid for this product, and your agreement to abide by the terms and conditions of this Agreement, the Company grants to you a nonexclusive right to use and display the copy of the enclosed software program (hereinafter the "SOFTWARE") on a single computer (i.e., with a single CPU) at a single location so long as you comply with the terms of this Agreement. The Company reserves all rights not expressly granted to you under this Agreement.

2. **OWNERSHIP OF SOFTWARE:** You own only the magnetic or physical media (the enclosed disks) on which the SOFTWARE is recorded or fixed, but the Company retains all the rights, title, and ownership to the SOFTWARE recorded on the original disk copy(ies) and all subsequent copies of the SOFTWARE, regardless of the form or media on which the original or other copies may exist. This license is not a sale of the original SOFTWARE or any copy to you.

3. **COPY RESTRICTIONS:** This SOFTWARE and the accompanying printed materials and user manual (the "Documentation") are the subject of copyright. You may <u>not</u> copy the Documentation or the SOFTWARE, except that you may make a single copy of the SOFTWARE for backup or archival purposes only. You may be held legally responsible for any copying or copyright infringement which is caused or encouraged by your failure to abide by the terms of this restriction.

4. **USE RESTRICTIONS:** You may <u>not</u> network the SOFTWARE or otherwise use it on more than one computer or computer terminal at the same time. You may physically transfer the SOFTWARE from one computer to another provided that the SOFTWARE is used on only one computer at a time. You may <u>not</u> distribute copies of the SOFTWARE or Documentation to others. You may <u>not</u> reverse engineer, disassemble, decompile, modify, adapt, translate, or create derivative works based on the SOFTWARE or the Documentation without the prior written consent of the Company.

5. **TRANSFER RESTRICTIONS:** The enclosed SOFTWARE is licensed only to you and may <u>not</u> be transferred to any one else without the prior written consent of the Company. Any unauthorized transfer of the SOFTWARE shall result in the immediate termination of this Agreement.

6. **TERMINATION:** This license is effective until terminated. This license will terminate automatically without notice from the Company and become null and void if you fail to comply with any provisions or limitations of this license. Upon termination, you shall destroy the Documentation and all copies of the SOFTWARE. All provisions of this Agreement as to warranties, limitation of liability, remedies or damages, and our ownership rights shall survive termination.

7. **MISCELLANEOUS:** This Agreement shall be construed in accordance with the laws of the United States of America and the State of New York and shall benefit the Company, its affiliates, and assignees.

8. **LIMITED WARRANTY AND DISCLAIMER OF WARRANTY:** The Company warrants that the SOFTWARE, when properly used in accordance with the Documentation, will operate in substantial conformity with the description of the SOFTWARE set forth in the Documentation. The Company does not warrant that the SOFTWARE will meet your requirements or that the operation of the SOFTWARE will be uninterrupted or error-free. The Company warrants that the media on which the SOFTWARE is

delivered shall be free from defects in materials and workmanship under normal use for a period of thirty (30) days from the date of your purchase. Your only remedy and the Company's only obligation under these limited warranties is, at the Company's option, return of the warranted item for a refund of any amounts paid by you or replacement of the item. Any replacement of SOFTWARE or media under the warranties shall not extend the original warranty period. The limited warranty set forth above shall not apply to any SOFTWARE which the Company determines in good faith has been subject to misuse, neglect, improper installation, repair, alteration, or damage by you. EXCEPT FOR THE EXPRESSED WARRANTIES SET FORTH ABOVE, THE COMPANY DISCLAIMS ALL WARRANTIES, EXPRESS OR IMPLIED, INCLUDING WITHOUT LIMITATION, THE IMPLIED WARRANTIES OF MERCHANTABILITY AND FITNESS FOR A PARTICULAR PURPOSE. EXCEPT FOR THE EXPRESS WARRANTY SET FORTH ABOVE, THE COMPANY DOES NOT WARRANT, GUARANTEE, OR MAKE ANY REPRESENTATION REGARDING THE USE OR THE RESULTS OF THE USE OF THE SOFTWARE IN TERMS OF ITS CORRECTNESS, ACCURACY, RELIABILITY, CURRENTNESS, OR OTHERWISE.

IN NO EVENT, SHALL THE COMPANY OR ITS EMPLOYEES, AGENTS, SUPPLIERS, OR CONTRACTORS BE LIABLE FOR ANY INCIDENTAL, INDIRECT, SPECIAL, OR CONSEQUENTIAL DAMAGES ARISING OUT OF OR IN CONNECTION WITH THE LICENSE GRANTED UNDER THIS AGREEMENT, OR FOR LOSS OF USE, LOSS OF DATA, LOSS OF INCOME OR PROFIT, OR OTHER LOSSES, SUSTAINED AS A RESULT OF INJURY TO ANY PERSON, OR LOSS OF OR DAMAGE TO PROPERTY, OR CLAIMS OF THIRD PARTIES, EVEN IF THE COMPANY OR AN AUTHORIZED REPRESENTATIVE OF THE COMPANY HAS BEEN ADVISED OF THE POSSIBILITY OF SUCH DAMAGES. IN NO EVENT SHALL LIABILITY OF THE COMPANY FOR DAMAGES WITH RESPECT TO THE SOFTWARE EXCEED THE AMOUNTS ACTUALLY PAID BY YOU, IF ANY, FOR THE SOFTWARE.

SOME JURISDICTIONS DO NOT ALLOW THE LIMITATION OF IMPLIED WARRANTIES OR LIABILITY FOR INCIDENTAL, INDIRECT, SPECIAL, OR CONSEQUENTIAL DAMAGES, SO THE ABOVE LIMITATIONS MAY NOT ALWAYS APPLY. THE WARRANTIES IN THIS AGREEMENT GIVE YOU SPECIFIC LEGAL RIGHTS AND YOU MAY ALSO HAVE OTHER RIGHTS WHICH VARY IN ACCORDANCE WITH LOCAL LAW.

ACKNOWLEDGMENT

YOU ACKNOWLEDGE THAT YOU HAVE READ THIS AGREEMENT, UNDERSTAND IT, AND AGREE TO BE BOUND BY ITS TERMS AND CONDITIONS. YOU ALSO AGREE THAT THIS AGREEMENT IS THE COMPLETE AND EXCLUSIVE STATEMENT OF THE AGREEMENT BETWEEN YOU AND THE COMPANY AND SUPERSEDES ALL PROPOSALS OR PRIOR AGREEMENTS, ORAL, OR WRITTEN, AND ANY OTHER COMMUNICATIONS BETWEEN YOU AND THE COMPANY OR ANY REPRESENTATIVE OF THE COMPANY RELATING TO THE SUBJECT MATTER OF THIS AGREEMENT.

Should you have any questions concerning this Agreement or if you wish to contact the Company for any reason, please contact in writing at the address below.

Robin Short
Prentice Hall PTR
One Lake Street
Upper Saddle River, New Jersey 07458